T0192827

SpringerBriefs in Applied Sciences and Technology

Nanoscience and Nanotechnology

Indexed by SCOPUS Nanoscience and nanotechnology offer means to assemble and study superstructures, composed of nanocomponents such as nanocrystals and biomolecules, exhibiting interesting unique properties. Also, nanoscience and nanotechnology enable ways to make and explore design-based artificial structures that do not exist in nature such as metamaterials and metasurfaces. Furthermore, nanoscience and nanotechnology allow us to make and understand tightly confined quasi-zero-dimensional to two-dimensional quantum structures such as nanopalettes and graphene with unique electronic structures. For example, today by using a biomolecular linker, one can assemble crystalline nanoparticles and nanowires into complex surfaces or composite structures with new electronic and optical properties. The unique properties of these superstructures result from the chemical composition and physical arrangement of such nanocomponents (e.g., semiconductor nanocrystals, metal nanoparticles, and biomolecules). Interactions between these elements (donor and acceptor) may further enhance such properties of the resulting hybrid superstructures. One of the important mechanisms is excitonics (enabled through energy transfer of exciton-exciton coupling) and another one is plasmonics (enabled by plasmon-exciton coupling). Also, in such nanoengineered structures, the light-material interactions at the nanoscale can be modified and enhanced, giving rise to nanophotonic effects.

These emerging topics of energy transfer, plasmonics, metastructuring and the like have now reached a level of wide-scale use and popularity that they are no longer the topics of a specialist, but now span the interests of all "end-users" of the new findings in these topics including those parties in biology, medicine, materials science and engineerings. Many technical books and reports have been published on individual topics in the specialized fields, and the existing literature have been typically written in a specialized manner for those in the field of interest (e.g., for only the physicists, only the chemists, etc.). However, currently there is no brief series available, which covers these topics in a way uniting all fields of interest including physics, chemistry, material science, biology, medicine, engineering, and the others.

The proposed new series in "Nanoscience and Nanotechnology" uniquely supports this cross-sectional platform spanning all of these fields. The proposed briefs series is intended to target a diverse readership and to serve as an important reference for both the specialized and general audience. This is not possible to achieve under the series of an engineering field (for example, electrical engineering) or under the series of a technical field (for example, physics and applied physics), which would have been very intimidating for biologists, medical doctors, materials scientists, etc.

The Briefs in NANOSCIENCE AND NANOTECHNOLOGY thus offers a great potential by itself, which will be interesting both for the specialists and the non-specialists.

More information about this subseries at https://link.springer.com/bookseries/11713

Yong Kang Eugene Tay · Huajun He ·
Xiangling Tian · Mingjie Li · Tze Chien Sum

Halide Perovskite Lasers

Yong Kang Eugene Tay
School of Physical and Mathematical Sciences
Nanyang Technological University
Singapore, Singapore

Xiangling Tian
School of Physical and Mathematical Sciences
Nanyang Technological University
Singapore, Singapore

Tze Chien Sum
School of Physical and Mathematical Science
Nanyang Technological University
Singapore, Singapore

Huajun He
School of Physical and Mathematical Sciences
Nanyang Technological University
Singapore, Singapore

Mingjie Li
Department of Applied Physics
The Hong Kong Polytechnic University
Hung Hom, Hong Kong

ISSN 2191-530X ISSN 2191-5318 (electronic)
SpringerBriefs in Applied Sciences and Technology
ISSN 2196-1670 ISSN 2196-1689 (electronic)
Nanoscience and Nanotechnology
ISBN 978-981-16-7972-8 ISBN 978-981-16-7973-5 (eBook)
https://doi.org/10.1007/978-981-16-7973-5

This Springer imprint is published by the registered company Springer Nature Singapore Pte Ltd.
The registered company address is: 152 Beach Road, #21-01/04 Gateway East, Singapore 189721, Singapore

Preface

Due to the increasing attention and appeal of Halide Perovskites as the next-generation solar cell and light-emitting sources, the compilation of content in this book is designed to cater for beginner post-graduate students and/or post-doctorates intending to work on the field of Halide Perovskite lasers. Therefore, the objective of this book is to serve as an introductory level reference book covering a wide range of crucial concepts from the spectroscopic characterisation of laser emissions to the properties and experimental results of Halide Perovskite materials for lasing studies. We strongly believe that with the availability and accessibility of this reference book, researchers in this field can easily and quickly "come up to speed" with the updates in this community. In order to maximise the effectiveness of this book, readers are expected to have background knowledge of ray-optics, wave-optics, quantum mechanics and integral calculus. It is important to acknowledge that basic calculus is a fundamental foundation behind the theoretical framework of laser physics and thus, is necessary in order to translate from mathematical to physical and intuitive understanding of the arguments involved. By further considering that researchers (such as chemists and material scientists) intending to enter this field may not be well-equipped with elementary laser physics (which is extremely crucial for a well-rounded understanding of this field), the introduction sections presented in Chap. 1 especially are written in a comprehensive yet concise manner. The intended outcome is to provide the readers with necessary exposure to both the technical jargons and underlying science behind laser characterisations. Certainly, readers who wish to expand on the knowledge of lasers are encouraged to look up on well-established reference books found in the bibliography.

In this book, the contents are arranged such that Chap. 1 covers the fundamentals and various ways to characterise laser emissions. Chapter 2 aims to introduce the Perovskite semiconducting materials and its basic morphologies and properties. Chapter 3 outlines the photo-physics of the underlying mechanisms leading to observations of optical gain in Perovskites. Specifically, we distinguish between free-carrier and exciton carrier dynamics in different tri-halide systems. Finally, in Chap. 4, we showcase both the research milestones and the current state-of-the-art system records achieved up till year 2019. For a clearer view of the feats achieved, we shall present them based on its material morphology as they hold key to the type of lasing observed. Primarily, we are concerned with physical scaling bulk single-crystals to the nanocrystalline morphologies and its viability as lasing gain materials. Furthermore, we will also dedicate a small discussion on a closely related phenomenon to photon lasing, called polariton lasing. Lastly, we share some perspectives on the direction of this research field by analysing the remaining challenges to be tackled, with the ultimate goal of realising low cost and performance stable electrically driven Perovskite lasers.

Singapore, Singapore — Yong Kang Eugene Tay
Singapore, Singapore — Huajun He
Singapore, Singapore — Xiangling Tian
Hung Hom, Hong Kong — Mingjie Li
Singapore, Singapore — Tze Chien Sum

Contents

Chapter 1
Introduction: Fundamentals of Lasers

1.1 What is a Laser?

The term LASER is an acronym for Light Amplification by Stimulated Emission of Radiation [1]. In today's world, the Laser is a convenient handheld device that produces ***coherent*** light and has found widespread industrial and scientific applications. For instance, infrared lasers are implemented in barcode scanners [2], in thermometers for concreting [3] and in laser levels for site survey [4]. Other applications include dentistry [5], ophthalmology [6], optical communications [7], industrial sensory [8] and data storage applications [9]. The first working laser was the red-emitting (694.3 nm) ruby laser constructed by Theodore Maiman on 16th May 1960 [10], as shown in Fig. 1.1.

This success spurred the expansion of alternative gain materials, such as the well-known Nd: YAG (solid state), Rhodamine6G (dye) and He–Ne (gas) lasers, all operating in the visible spectrum [12]. The construction of a laser requires three components: a gain medium, a pump source and an external resonator or cavity enclosing the gain medium, which will be discussed later. Section 1.2 explores the role of a gain medium, by considering the interplay between radiative processes under pumping conditions. Next, Sect. 1.3 introduces the third component: optical feedback, where we shall see that the provision of optical feedback from cavity couples to the pumped gain medium effectively to produce laser output.

1.2 The Gain Medium

In this section, we consider the first component of a laser system: the gain medium. Essentially, the gain medium is a material responsible for amplifying light through stimulated emission with optical coherence properties. We begin Sect. 1.2.1 by reviewing various radiative transitions occurring in an arbitrary two-level atomic system. Section 1.2.2 discusses how stimulated emission under a condition called

Y. K. E. Tay et al., *Halide Perovskite Lasers*, Nanoscience and Nanotechnology,
https://doi.org/10.1007/978-981-16-7973-5_1

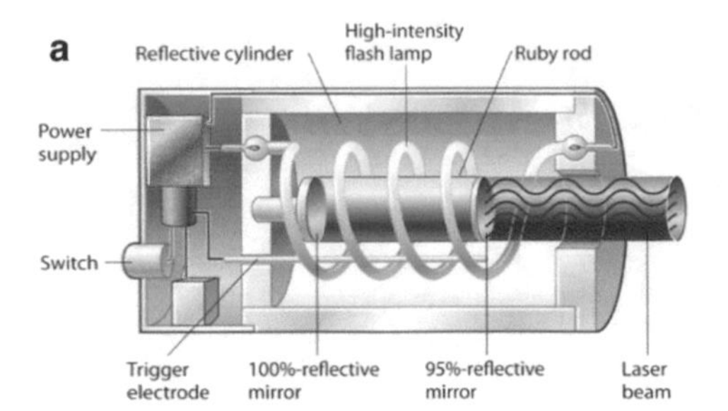

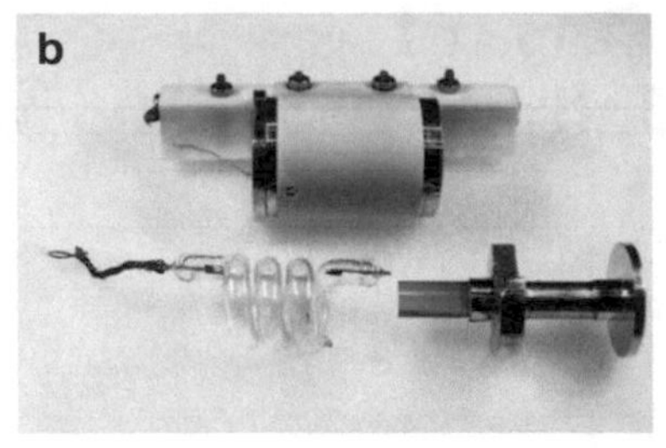

Fig. 1.1 The **a** schematic of an assembled ruby laser constructed in 1960 by Theodore Maiman at Hughes research laboratories and **b** its three major components disassembled, showing a spiral-shaped flash lamp acting as the pump source, a cylindrical holder with reflective interior for providing optical feedback to the gain medium's light output and a red ruby rod acting as the gain material [11]

population inversion is a prerequisite for light amplification, which in turn is a prerequisite for lasing. As we shall see, the ease of population inversion largely depends on the material's energy level landscape and determines the material's gain coefficient, which will be discussed in Sect. 1.2.3. Also, due to energy conservation laws, we further account for inevitable gain saturation. Finally, Sect. 1.2.4 introduces the concept of Amplified spontaneous Emission (ASE), which is a manifestation of optically amplified light in the gain medium in the absence of optical feedback.

1.2.1 Radiative Transition Rates and Einstein's Relations

The radiative processes occurring in a medium is best explained using a simplistic two-level approach of an arbitrary atomic system in thermal equilibrium. To quantify the magnitudes of these following radiative processes, we shall introduce the Einstein's coefficients [1]. Initially, as shown in Fig. 1.2a, all atoms in the system occupies the ground state E_0, as the absence of a pump means no energy is supplied to the system. Now, when the pump is turned on, conservation of energy dictates that the atoms in ground state E_0 to absorb the pump energy and transit to the excited state

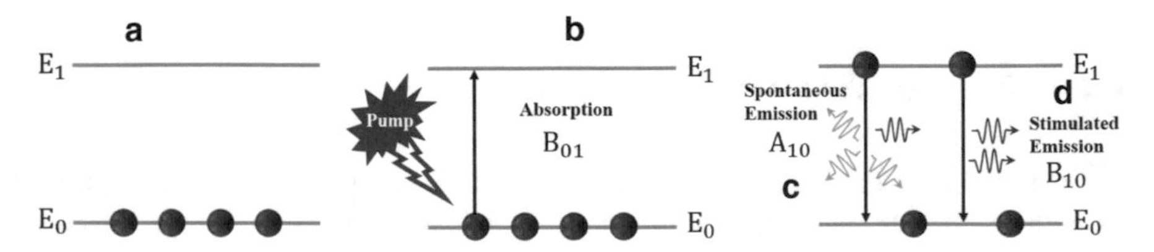

Fig. 1.2 **a** In the absence of pump, all carriers (Red spheres) reside in the lowest energy state E_0. **b** Pump absorption with **c** Spontaneous and **d** Stimulated emission processes

E_1, as shown by the upward $E_0 \rightarrow E_1$ transition arrow in Fig. 1.2b. Since absorption can only occur through external stimulus inputs, it is also conceptually known as stimulated absorption and is quantified by the coefficient B_{01}. In laser action, we are also concerned with a second kind of absorption known as re-absorption, which will be discussed later in the Sect. 1.2.2. Soon after the stimulated absorption event (typically several ns), excited atoms in state E_1 will relax back to its ground state E_0 either radiatively or non-radiatively (kinetic or internal decay). In the case of radiative decay, the excited atom may relax either by (I) Spontaneous Emission or (II) Stimulated Emission, as shown in Fig. 1.2c, d, respectively. Although both downward transition emit photons with energy $\hbar\omega = E_1 - E_0$, the two emission processes differ in terms of the photon properties. Spontaneous emission is a ***naturally occurring*** decay that emit photons with random wave phase and direction and is quantified by the coefficient A_{10}. On the other hand, stimulated emission requires an incident propagating photon to induce the downward transition that produces another "copy" photon with similar wave traits. Hence, stimulated emission is considered as the exact opposite process to (stimulated) absorption and is quantified by the coefficient B_{10}. Importantly, the onset of stimulated emission can be triggered when spontaneously emitted photons act as ***seed*** propagating photons to induce and create more photons with similar wave traits necessary for lasing action [1, 12].

Given that both stimulated absorption and emission are related, and that stimulated emission depends on the contribution from spontaneous emission, we realise that B_{01}, A_{10} and B_{10} are mutually dependent, which is the basis of the two Einstein's Relations. The *first Einstein's relation* states that stimulated absorption and emission are equal and exact opposite processes, whose radiative coefficients are expressed as:

$$g_0 B_{01} = g_1 B_{10} \tag{1.1}$$

where g_0 and g_1 are the degeneracy factors of energy states E_0 and E_1, respectively. For semiconductors with degenerate carrier states within the band structure, the first Einstein's relation can be written as $g_V B_{VC} = g_C B_{CV}$, where g_C and g_V are the conduction and valence band degeneracies where the stimulated emission occurred. The *second Einstein's relation* states that the two emission coefficients are dependent despite their intrinsic differences in emitted photon properties, and is given by [1]:

$$\frac{A_{10}}{B_{10}} = \frac{8\pi h\nu^3}{c^3} \tag{1.2}$$

where, $B_{10} \propto A_{10}$ implies that a strong stimulated emission rate must be assisted by a strongly-seeding spontaneous emission. It is important to note that the spontaneous (A_{10}) and stimulated (B_{ij}) coefficients have different units. For spontaneous emission, $[A_{10}] \equiv \mathrm{s}^{-1}$ is quantified by probability per unit time while stimulated process have units of $\left[B_{ij}\right] \equiv \mathrm{J}^{-1}\,\mathrm{m}^3\,\mathrm{s}^{-2}$ due to the reliance of external energy stimuli. In defining transition rates to be the probability of transition per unit time, we see that

A_{10} coefficient is exactly the spontaneous emission rate itself, but the stimulated absorption W_{01} and emission W_{10} transition rates are generally defined as:

$$W_{ij} = B_{ij}\rho_{\nu} \quad (1.3)$$

where W_{ij} depends on the input pump's spectral energy density ρ_{ν} (with units of J m^{-3} s) and that $W_{ij} = W_{ji}$ by virtue of Eq. (1.1). In most spectroscopic studies using steady-state photoluminescence (PL), the PL spectrum is associated with the material's spontaneous emission, with a typical fluorescence lifetime estimated as $\tau_{PL} = \frac{1}{A_{10}}$. In Sect. 1.2.2, we shall consider the various energy level models that can aid in promoting stimulated emission that is crucial for achieving light amplification.

1.2.2 Population Inversion in Two, Three and Four-Level Systems

In laser operation, the role of the gain medium is to amplify light by the process of stimulated emission. In other words, a stream of incident photons entering a gain medium will exit with even more photons. As such, it is important for stimulated emission contributions to exceed absorption losses and we shall consider two kinds of absorption: (i) pump (stimulated) absorption and (ii) reabsorption of emitted photons. For this reason, we turn to a rate equation approach in order to assess the condition necessary to attain net optical gain.

Starting with a simple *two-level system* shown in Fig. 1.3a, since the system only comprises of a ground E_0 and immediate excited state E_1, the stimulated absorption and emission occur along the same transition, where $W_{abs} = W_{SE}$. If the rate of increase in photon flux due to stimulated emission is given by $\left(\frac{d\phi}{dt}\right)_{SE} \propto N_{up}W_{SE}$ while the decrease in photon flux due to absorption is given by $\left(\frac{d\phi}{dt}\right)_{abs} \propto N_{low}W_{abs}$, then, the net rate of change of photon flux $\left(\frac{d\phi}{dt}\right)_{net} = \left(\frac{d\phi}{dt}\right)_{SE} - \left(\frac{d\phi}{dt}\right)_{abs}$ is explicitly given by:

$$\left(\frac{d\phi}{dt}\right)_{net}^{2L} \propto \Delta N^{2L} W_{SE} \quad (1.4)$$

where $\Delta N = N_1 - N_0$ is the **population difference** between the upper and lower energy state for which the stimulated emission transition takes place. Since optical gain corresponds to $\left(\frac{d\phi}{dt}\right)_{net} > 0$, it means that $\Delta N > 0$, where $N_1 > N_0$. This "abnormal" condition where the upper excited state occupancy is greater than its lower state occupancy is known as **population inversion**. Unfortunately, population inversion **cannot** occur in two-level systems because of two major reasons.

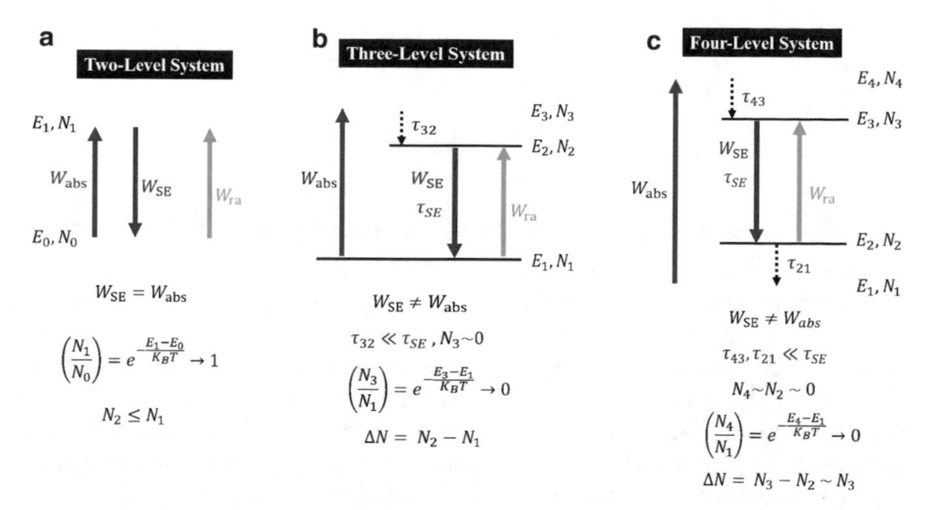

Fig. 1.3 A comparison between **a** two-level, **b** three-level and **c** four-level systems. Dotted lines indicate quick non-radiative decay processes. Blue, red, and gray lines represent Stimulated absorption (abs), emission (SE) and reabsorption (ra) transitions, respectively

Firstly, since the stimulated emission rate is exactly countered by its absorption ($W_{abs} = W_{SE}$), there is no net gain. Secondly, from a statistical mechanics viewpoint, the *Boltzmann factor* governing the state occupancies under thermal equilibrium conditions dictates that $\left(\frac{N_1}{N_0}\right) = e^{-\frac{h\nu}{K_B T}}$, which forbids $N_1 > N_0$. In practice, another loss channel known as reabsorption should also be included, which further inhibits the development of optical gain. A quintessential example is the single exciton gain mechanism discussed in Chap. 3, where the exact compensation between excitonic absorption and stimulated emission transitions hinders population inversion of primary exciton carriers.

The minimum requirement is that the gain medium must have a *three-level system* shown in Fig. 1.3b. Here, the solution for successful population inversion is due to the presence of a longer-lived **metastable** state E_2 that is used to not only accumulate excited atoms till $N_2 > N_1$ is met, but also decouple the stimulated absorption and emission transition lines. As such, the single Boltzmann factor limitation is nullified because the absorption and stimulated emission transitions are different. Instead, stimulated absorption occurs along $E_1 \rightarrow E_3$ with rate $W_{abs} = W_{31}$ while stimulated emission occurs along $E_2 \rightarrow E_1$ with rate $W_{SE} = W_{21}$ [13]. By assuming that $E_3 \rightarrow E_2$ is a quick non-radiative thermalisation, we may approximate $N_3 \sim 0$. As such, we notice that the Boltzmann factors governing the pump absorption and stimulated emission transitions are given by $\left(\frac{N_3}{N_1}\right) \rightarrow 0$ and $\left(\frac{N_2}{N_1}\right) = e^{-\frac{E_2 - E_1}{K_B T}}$, respectively. Along the stimulated emission transition line, we can write the net change in photon flux density and population inversion terms as:

$$\left(\frac{d\phi}{dt}\right)_{net}^{3L} = \left(\frac{d\phi}{dt}\right)_{SE} - \left(\frac{d\phi}{dt}\right)_{ra} = \Delta N^{3L} W_{21} \quad (1.5)$$

$$\Delta N^{3L} = N_2 - N_1 \quad (1.6)$$

where $\left(\frac{d\phi}{dt}\right)_{SE}$ is required to exceed the reabsorption losses of $\left(\frac{d\phi}{dt}\right)_{ra}$ of emitted photons with rate $W_{12} = W_{21}$ by virtue of the first Einstein's relation [13]. While the three-level system is just sufficient to achieve optical gain, its pump-inefficiency is a major drawback [1, 12]. This is because the ground state involvement as the lower level indicate that the gain material must be pumped intensely, such that more than half of the ground state occupancy is excited to E_2 [1, 12]. A quintessential example is the biexciton gain mechanism in green-emitting inorganic Cesium Lead Bromide ($CsPbBr_3$) Perovskites, where stimulated emission occurs between a secondary biexciton state and a primary free exciton state.

Lastly, in a ***four-level system*** as shown in Fig. 1.3c, stimulated emission occurs along $E_3 \rightarrow E_2$. Here, the lower level does not involve the ground-state E_1, thereby circumventing the requirement to pump more than half of the atoms up to the excited states and makes them extremely efficient. While the pump occurs along $E_1 \rightarrow E_4$, the fast non-radiative decay $\tau_{43} \ll \tau_{SE}$ causes excited atoms to quickly populate the metastable E_3 for accumulating population inversion. As such, $N_4 \sim 0$ and so $\left(\frac{N_4}{N_1}\right) \rightarrow 0$. Since the stimulated emission leading to optical gain occurs along $E_3 \rightarrow E_2$, the net rate of change to the photon flux density and population difference term can be written analogously:

$$\left(\frac{d\phi}{dt}\right)_{net}^{4L} = \left(\frac{d\phi}{dt}\right)_{SE} - \left(\frac{d\phi}{dt}\right)_{ra} = \Delta N^{4L} W_{32} \quad (1.7)$$

$$\Delta N^{4L} = N_3 - N_2 \approx N_3 \quad (1.8)$$

where $W_{23} = W_{32}$ by virtue of the first Einstein's relation [13]. If the lower level E_2 is also a short-lived state ($\tau_{21} \ll \tau_{SE}$) where atoms quickly decay to E_1, $N_2 \sim 0$ and population inversion condition of $N_3 > N_2$ is easily satisfied even if N_3 is small at very weak pump regimes. Thus, we expect optical gain arising from four-level systems to possess lower pumping thresholds than three-level systems.

1.2.3 The Gain Coefficient and Gain Saturation

The magnitude of optical gain in a material is quantified by the gain coefficient g and is defined as the fractional increase of light intensity per unit distance: $g \equiv \frac{\left(\frac{dI}{I}\right)}{dx}$ with

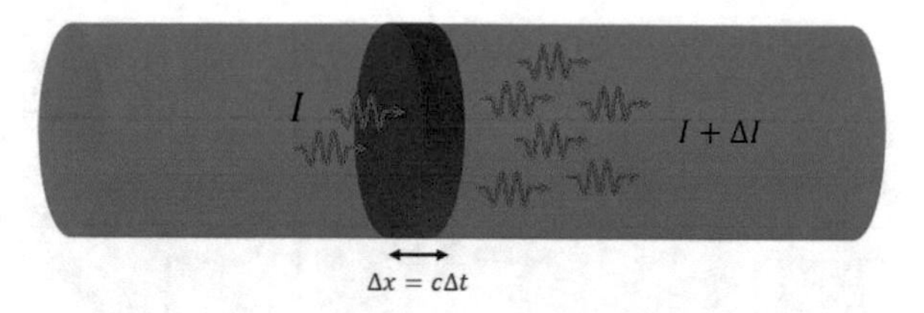

Fig. 1.4 A schematic of optical gain causing light intensity to increase from I to I + ΔI as photons propagates over Δx in the gain medium. Here, stimulated emission occurs under population inversion in the medium

units of cm^{-1}. For simplicity, we consider a one-dimensional amplifier in Fig. 1.4, where stimulated emission photons traverse through an infinitesimal distance Δx.

The rate of change of light intensity is therefore given by [14]:

$$\frac{dI}{dx} = A + gI - \alpha I = A + g_{net} I \tag{1.9}$$

where A, g and α are the spontaneous emission seed, intrinsic material optical gain and loss (propagation and re-absorption) coefficients, respectively and that $g_{net} = g - \alpha$ is the resulting net gain after overcoming losses. Solving Eq. (1.9) yields the result [15]:

$$I(x) = \frac{A}{g_{net}}\left(e^{g_{net}x} - 1\right) \tag{1.10}$$

where Eq. (1.10) predicts the light intensity to build up exponentially with propagated distance in the gain medium. However, it also predicts $I(x) \to \infty$ if $x \to \infty$. In practice, the material is not infinitely long and light does not build up infinitely under a constant pumping rate because otherwise, energy conservation would be violated [16]. Thus, this exponential increase in light intensity is termed as the **unsaturated gain**, describing the initial situation (say $x = 0$), where input light intensity is low (low signal gain [1, 12]). As photons propagate further, say $x > x_s$, this exponential rate of intensity growth will start to decrease and eventually, light intensity approach I_{max} asymptotically. This phenomenon is known as **gain saturation** and occurs due finite maximum inversion of carriers and the rapid depopulation of the metastable level occupancies by the highly intense input light. By employing reverse Taylor's expansion to the unsaturated gain solution in Eq. (1.10), one obtains the gain saturation accounted solution given by [17]:

$$I(x) = A\left[e^{g_{net}x_s\left(1-e^{-x/x_s}\right)} - 1\right] \tag{1.11}$$

where x_s is defined as the saturation distance. Here, an initial weak input light is expected to grow exponentially under the unsaturated gain regime at $x \leq x_s$ but

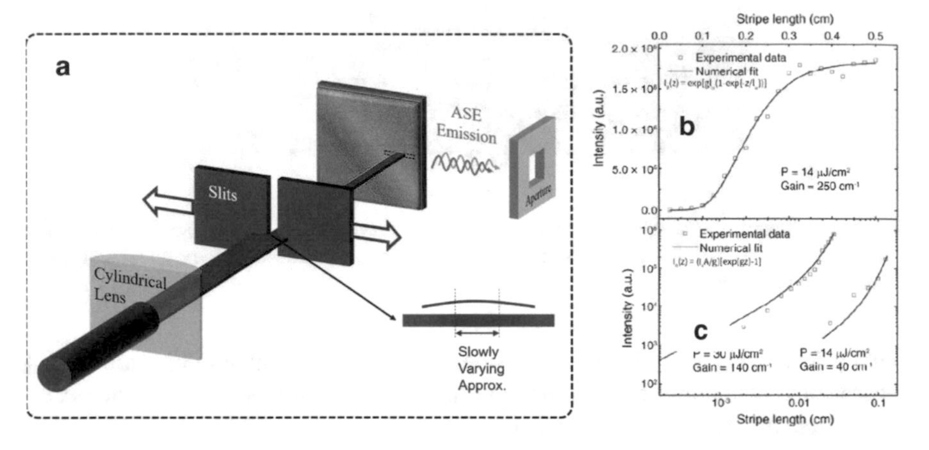

Fig. 1.5 **a** A schematic of the Variable Stripe-Length (VSL) experiment commonly used to determine the gain coefficients of a gain material. An example of gain-curve fitting using the **b** saturation accounted and **c** unsaturated gain solutions described by Eqs. (1.12) and (1.11) respectively

is also expected to pick up lesser intensity increment as $x \geq x_s$ as a result of gain saturation, tending asymptotically towards $e^{g_{net}x_s}$. At exactly $x = x_s$, we have a factor of $(1 - e^{-1})$, which implies that the optical buildup is near its maximum attainable value, with only e^{-1} remaining proportion left to grow before reaching the maximum value of $e^{g_{net}x_s}$. In optical spectroscopic studies evaluating the gain coefficient in a Perovskite media, the so-called Variable Stripe-Length (VSL) geometry [15] shown in Fig. 1.5a is often used. Developed by Shaklee and Leheny, VSL is a convenient method as the net gain coefficient of a material can be determined directly without having to construct cumbersome laser cavity systems [15].

Here, a gaussian pump beam is focused with a cylindrical lens to produce a line beam that is illuminated onto the Perovskite medium. The beam stripe length x is controlled by a pair of mechanical slits. To ensure a homogeneous excitation, slits are usually placed at the center of the stripe excitation beam, based on the slowly varying approximation in gaussian beam optics [10]. Figure 1.5b, c shows the gain coefficient fitting of methylammonium lead iodide ($CH_3NH_3PbI_3$) thin films [18] collected from the VSL experiments using the saturation accounted and unsaturated gain solutions. Clearly, the saturation accounted solution in Eq. (1.11) provides a complete and more accurate fitting (Fig. 1.5b) than the unsaturated gain solution (Fig. 1.5c) as the latter only provided partial data fitting corresponding to shorter excited stripe lengths.

In most conventional laser physics reference books, gain saturation in atomic systems are understood through the relationship between gain coefficient, light intensity, and population inversion quantities in the framework of four-level systems, as shown in Fig. 1.6a. By including the spontaneous emission terms of A_{41} and A_{32}, the rate equation describing the population inversion dynamics can be written as:

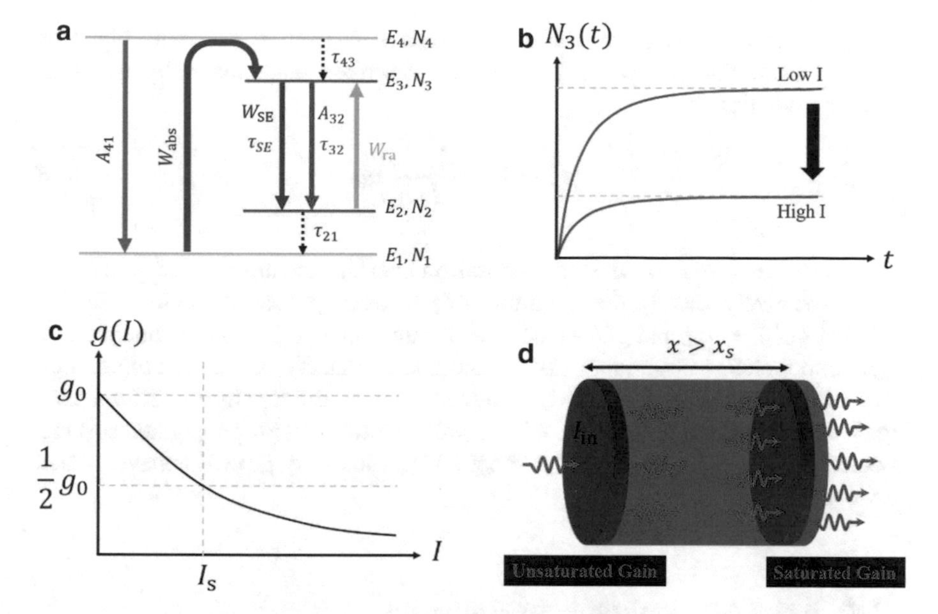

Fig. 1.6 **a** Radiative processes in a four-level system. Here, reabsorption rate W_{ra} is negligible due to fast τ_{32} making $N_2 \sim 0$. **b** Schematic illustrating the initial unsaturated and subsequent saturated gain with propagated distance x in the gain medium. **c** Maximum steady-state population inversion $N_3(\infty)$ that reduces as a consequence of **d** gain reduction with increasing light intensity

$$\frac{dN_3}{dt} = P + [-W_{32}N_3] + [-A_{32}N_3] \tag{1.12}$$

where $P = W_{41}N_1$ is defined as the pumping rate and that $W_{23}N_2 = A_{41}N_4 = 0$ since $N_2 \sim N_4 \approx 0$. By further defining the effective lifetime of upper level E_3 to be $\frac{1}{\tau_{3(\text{eff})}} = W_{32} + A_{32} \propto I$, we can re-write the above rate equation as follows:

$$\frac{dN_3}{dt} = P - \frac{N_3}{\tau_{3(\text{eff})}} \tag{1.13}$$

Solving Eq. (1.13) with boundary conditions of $N_3(0) = 0$ under lossless conditions gives the explicit time-dependent and steady-state upper level E_3 occupancies as follows:

$$N_3(t) = N_3(\infty)\left(1 - e^{-t/\tau_{3(\text{eff})}}\right) \tag{1.14}$$

$$N_3(\infty) = P\tau_{3(\text{eff})} \tag{1.15}$$

where $N_3(\infty) \propto P$ is the steady-state population inversion and is proportional to the pumping rate P. Equation (1.14) tells us that there is a characteristic buildup time leading to the steady-state population inversion of $N_3(\infty)$, as shown in Fig. 1.6b. Next, the saturation of optical gain can be described mathematically using the saturation function:

$$g(I) \approx \left(\frac{1}{1 + I/I_s}\right) g_0 \tag{1.16}$$

where I_s and $g_0 = g(I \to 0)$ are the saturation intensity and unsaturated gain coefficient, respectively. Clearly, the magnitude of gain exactly at saturation intensity I_s is such that $g(I_s) = \frac{1}{2} g_0$ and $g(I \to \infty) \to 0$. Since optical gain is dependent on the extent of population inversion, $g_0 \propto N_3$ and tells us that $N_3(\infty)$ at high input intensities will be significantly lowered, as shown by the red-curve in Fig. 1.6b. Lastly, Fig. 1.6d summarises the relationship of gain saturation with propagated distance x, corroborating to the expected "S-shaped" light intensity growth behaviour seen previously in Fig. 1.5b.

1.2.4 Amplified Spontaneous Emission

Thus far, we have seen that population inversion in the gain medium is a prerequisite pump-induced buildup for avalanche stimulated emission to occur and drives the onset of optical gain. Under such circumstances, optical gain unsupported by optical feedback results in amplified spontaneous emission (ASE) instead of lasing. As shown in Fig. 1.7, ASE results from the total contributions of various avalanche stimulated emission processes seeded by spontaneously emitted photons situated

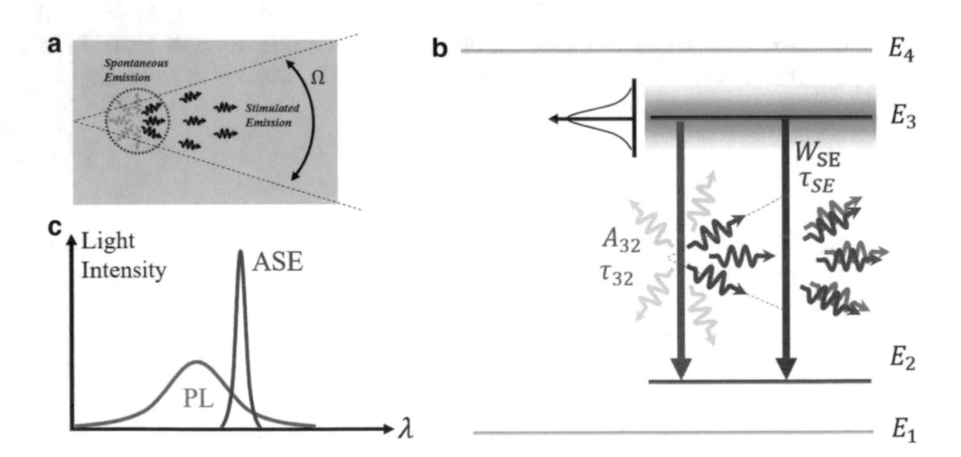

Fig. 1.7 **a** Amplified Spontaneous Emission (ASE) without optical feedback in a gain material occurring within the locus of Ω. **b** Schematic of ASE in a four-level system gain medium and **c** the development of narrowband ASE spectrum within the spontaneous emission PL spectrum

within the locus of a solid angle of $\Omega = \frac{\pi D^2}{4L^2}$, accumulated in a one-pass propagation [12]. In other words, ASE is a "less stringent" light amplification process where spontaneous emission is amplified via stimulated emission with "relaxed" optical coherence properties in comparison to lasers [12, 19]. In spectroscopic terms, the gain spectrum/bandwidth refers to the ASE spectrum (Fig. 1.7c, orange) with much higher intensity and narrower full-width at half maximum (FWHM) than its spontaneous emission PL spectral widths (green). Essentially, ASE possessing narrower spectral widths than its PL counterpart is a consequence of partial temporal coherence, which we shall re-visit in Sect. 1.4.1. Readers interested in the quantitative description of ASE are encouraged to refer to reference [12, 20].

1.3 Laser Oscillations: Gain Medium with Optical Feedback

Optical feedback refers to the wave-guiding of stimulated emission photons by re-directing light back and forth within the gain medium and can be easily achieved by placing two end-mirrors sandwiching the gain medium (similar to Fig. 1.1a). As such, light amplification no longer occurs in a "one-pass" propagation but "multipass", with one end of the mirror (output coupler) emitting some as laser output. Essentially, the waveguiding and confinement effect caused by the end-reflective mirrors *selectively* enhance specific avalanched photons called modes while annihilating other avalanched photons (non-modes), revealing more stringent light amplification conditions than ASE. To illustrate this, Sect. 1.3.1 discusses basic resonator optics, showing that the selective light amplification producing laser emission is a result of mathematical boundary conditions due to the spatial confinement. Section 1.3.2 illustrates the features of steady-state laser output and lastly, Sect. 1.3.3 discuss the various figures of merit characterising the quality of modal confinement.

1.3.1 Cavity Modes and Lasing Modes

In this section, we will use the framework of Fabry–Perot cavities to illustrate the concept of cavity modes, gain spectrum and lasing modes. From Fig. 1.8a, the presence of two end-reflective mirrors imposes the resonant boundary condition of $m\lambda \approx 2nL$, where only light that constructively interfere with itself after propagating a cavity round trip distance of $2nL$ can exist in the cavity with length L. Solving the boundary condition yields the solution:

$$\nu_m \approx m\left[\frac{c}{2nL}\right] \tag{1.17}$$

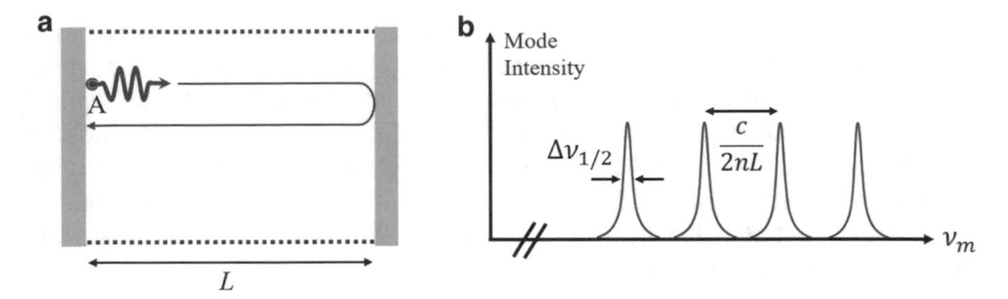

Fig. 1.8 **a** Light confinement in a Fabry–Perot resonator with length L. **b** Cavity mode frequency comb with narrow spectral widths of $\Delta\nu_{1/2}$ due to the photon's energy-time uncertainty relations

where ν_m is the collection of quantized cavity mode frequencies allowed in the cavity and m is an integer called the mode number—Eq. (1.17).

In practice, cavity mode frequencies are broadened to width $\Delta\nu_{1/2}$ instead of its dirac-sharp feature described in Eq. (1.18) because of the photon energy-lifetime uncertainty relations: $(h\Delta\nu_{1/2})\Delta\tau \sim \hbar$ originating from imperfect reflectivities R_1R_2. In other words, cavity mode widths are an indication of quality of optical confinement (feedback) and is defined by [1]:

$$\Delta\nu_{1/2} = \frac{1}{2\pi}[1 - R_1R_2]\left[\frac{c}{2nL}\right] \tag{1.18}$$

where the stronger the optical feedback $R_1R_2 \to 1$, the narrower $\Delta\nu_{1/2} \to 0$ will be. However, one must note that $R_1R_2 = 1$ is not exactly ideal too, as no output beam can be collected and leads to internal gain saturation due to high intensity build-ups [12]. In actual fabricated cavity systems with cavity length L, the experimentally observed cavity modes are situated spectrally within the PL band, as shown by the green modes in Fig. 1.9a. On the other hand, the lasing modes (orange) are defined

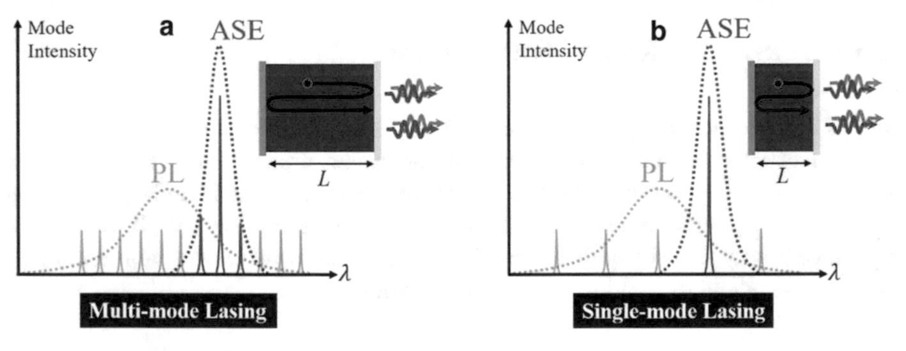

Fig. 1.9 Schematic illustrating the difference between **a** multi-mode and **b** single-mode lasing, which depends largely on the number of coinciding cavity modes with the ASE gain spectrum dictated by the FSR

as cavity modes found **only** within the ASE (or gain) spectrum that subsequently participate in laser oscillations and are responsible for the observed laser output wavelengths (or frequencies). As we will see in Sect. 1.3.2, cavity modes only evolve into lasing modes after the cavity system is pumped sufficiently above threshold, where population inversion leading to optical gain in the gain medium occurs.

One of the most important pursuits is ***single-mode lasing*** operation, due to benefits of strong spectral purity, preservation of optical coherence and excellent beam qualities [21, 22]. Generally, the number of participating lasing modes depends on the intermodal spacing between adjacent modes, called the free-spectral range (FSR). From Eq. (1.17), the FSR in frequency domain can be derived as follows:

$$\mathrm{FSR}_{\upsilon} = \nu_{m+1} - \nu_{m} = \left[\frac{c}{2nL}\right] \quad (1.19)$$

Clearly, the FSR is dependent on the gain medium's refractive index n and more importantly, the cavity length L. Essentially, a long cavity system typically causes small FSR which in turn leads to multimodal lasing (see Fig. 1.9a). Thus, single-mode laser outputs are usually achieved only when the FSR is large, so that there is minimal overlap between cavity modes and the gain spectrum, as shown in Fig. 1.9b. However, this also demands the cavity length L to be minimised, which can be an imposing limit on the gain medium thickness. Following in Sect. 1.3.2, we shall discuss the concept of laser oscillations, where beyond a certain threshold feature, pumping induced population inversion in the gain material couples with optical feedback to produce laser output. In Chap. 3, we shall see that the pursuit of single-mode lasers often involve leads to increased pumping threshold of lasing due to thinner active layers providing relatively lower optical gain per round-trip.

1.3.2 Steady-State Laser Oscillations and Gain Threshold Conditions

Laser oscillation refers to the situation where net gain in the medium is achieved, such that lasing mode intensities waveguided back and forth within the cavity are greatly amplified in comparison to non-participating cavity modes. For a cavity, the light intensity collected from point A to B after a round-trip is:

$$I_B = I_A\left[e^{g(2nL)}\right]\left[e^{-\alpha(2nL)}\right][R_1 R_2] \quad (1.20)$$

where we account for the intrinsic gain g, attenuation α and reflectivities $R_1 R_2$, respectively. To ensure net optical gain, we require $I_B > I_A$ but we can define gain threshold where intrinsic gain overcome losses exactly, such that $I_B = I_A$. This implies that the term $\left[e^{(g_{\mathrm{th}}-\alpha)(2nL)}\right][R_1 R_2] = 1$ and the gain coefficient threshold g_{th} can be solved to give:

$$g_{\text{th}} = \alpha + \frac{1}{2nL}\ln\left(\frac{1}{R_1R_2}\right) \propto N_{3(\text{th})} \tag{1.21}$$

Since $g \propto N_3$, the presence of g_{th} implies a corresponding $N_{3(\text{th})}$ exists. Equation (1.21) indicates that the mirrors used, and cavity length are important parameters during cavity system fabrications. In recalling Eq. (1.15), this also implies a corresponding pump threshold P_{th}. Essentially, g_{th} can be lowered through (I) Narrow gain bandwidth $\delta\nu$, (II) longer cavity length L, (III) small α and (IV) highly reflective $R_1R_2 \rightarrow 1$. Factors (I and II) describes the maximisation of intrinsic gain while (III, IV) minimises the collective losses. A narrower gain bandwidth helps to selectively maximise stimulated emission of a sharp transition line and reduce intermodal competition for oscillation, while a longer cavity length helps to boost the intrinsic round-trip gain, although it also hinders single-mode lasing operations and reduces compactness of the opto-electronic device. Thus, it is imperative to take note of the trade-offs involved for tailored cavity designs.

Equally important to laser oscillations, is the role of the pumping rate P, which can be addressed by considering steady-state laser oscillation as $t \rightarrow \infty$. Previously, our solution in Eq. (1.14) automatically describes the onset of optical gain as an ideal lossless four-level system has near-zero pumping threshold to accumulate population inversion. Here, to account for losses, where pump threshold P_{th} exists, we re-write the rate equation of population inversion as follows:

$$\left(\frac{dN_3(t)}{dt}\right) = P - \frac{N_3(t)}{\tau_{3(\text{eff})}} = 0 \tag{1.22}$$

where $\left(\frac{dN_3(t)}{dt}\right) = 0$ necessarily due to thermodynamic equilibrium of population inverted occupancies along the metastable state, where $N_3(\infty)$ is constant during steady-state conditions. Secondly, in assuming that the output coupler leaks out negligible laser output, the infinitely many round-trips of $x \rightarrow \infty$ would lead to saturation of gain, thereby maintaining a constant light intensity I_{max} internally, such that $\left(\frac{dI}{dt}\right) = 0$. In other words, regardless of pumping rate P, steady-state is said to be reached at $t \rightarrow \infty$ when both $\left(\frac{dN_3(t)}{dt}\right) = \left(\frac{dI}{dt}\right) = 0$ are simultaneously satisfied. At **below** threshold conditions of $P < P_{\text{th}}$, we expect the trivial solutions where (i) $P = \frac{N_3(\infty)}{\tau_{32}}$ and (ii) $I = 0$ due to no net gain, satisfying the two requirements, respectively. In (i), any increase in pumping rate $P < P_{th}$ will only serve to linearly increase $N_3(\infty) < N_{3(\text{th})}(\infty)$ that does not lead to any laser oscillations. At **threshold** conditions of $P = P_{\text{th}}$, we have $P_{\text{th}} = \frac{N_{3(\text{th})}(\infty)}{\tau_{32}}$ but $I = 0$ also, due to exactly zero net gain. Finally, at **above** threshold conditions of $P > P_{\text{th}}$, since $I \propto N_3(\infty)$, the requirement of $\left(\frac{dI}{dt}\right) = 0$ at steady-state operation dictates that $N_3(\infty) = N_{3(\text{th})}(\infty) \equiv N_{3(\text{max})}(\infty)$ necessarily, while lasing intensity increases linearly [1] with pump in this regime given by$I = \frac{N_3 h\nu}{I g_{net}\tau_{32}}\left[\frac{P}{P_{\text{th}}} - 1\right]$. Figure 1.10 summarises the transition in behaviours of steady-state population inversion $N_3(\infty)$ and lasing intensity I with respect to P at below to above pumping threshold regimes

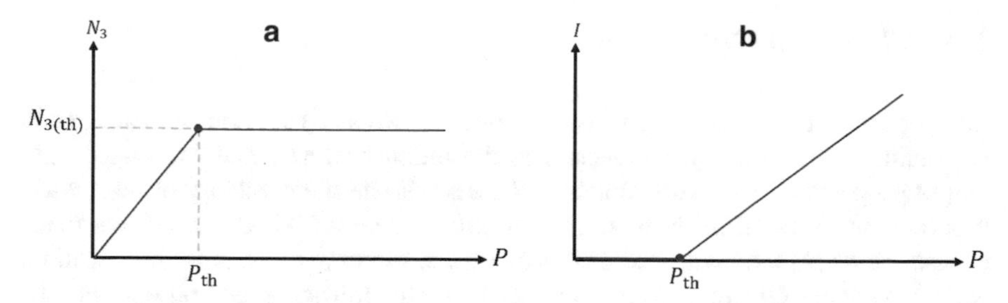

Fig. 1.10 The steady-state behaviour of laser oscillations. **a** Population inversion and **b** lasing light intensity as a function pumping rate P, with a threshold at $P = P_{th}$. Both shows a distinctive change in behaviour as $P \rightarrow P_{th}$

and are often plotted from experimental data to ascertain the onset of laser oscillations and to distinguish lasing from other "look-alike" narrowband emission artefacts [23], which we will discuss in Sect. 1.4.3.

1.3.3 Figures of Merit: Q-Factor and Finesse

Two common figures of merit used to characterise the output quality of a laser are the (I) Q-factor and (II) Finesse. As we have seen in Sect. 1.3.1, the mirror imperfections lead to a broadening of the cavity modes and hence the resulting lasing modes. Firstly, the Q-factor provides indication of the optical feedback based on the sharpness of lasing modes, given by:

$$Q = \frac{\nu_{las}}{\Delta\nu_{1/2}} = \frac{4\pi n L}{\lambda(1 - R_1 R_2)} \tag{1.23}$$

On the other hand, the cavity Finesse is often used to judge the spectral overlap between adjacent lasing modes by accounting for the FSR:

$$\mathcal{F} = \frac{\Delta\nu_{1/2}}{\text{FSR}} = \frac{2\pi}{(1 - R_1 R_2)} = \frac{Q}{m} \tag{1.24}$$

Evidently in both Eqs. (1.23) and (1.24), the better the cavity fabrication, the higher the values of Q and $\mathcal{F}$ because lasing mode width should be smaller than both lasing mode frequency and FSR distance ($\Delta\nu_{1/2} < \text{FSR} \ll \nu_{las}$). Interestingly, cavity Finesse is only dependent on $R_1 R_2$ and independent of cavity length L. In principle, the optical feedback is independent of L and should depend on the mirror $R_1 R_2$, therefore, a good laser cavity is one where the lasing modes are narrow and do not overlap with adjacent modes.

1.4 Characteristics of Lasers

Lasers possess two kinds of optical coherence, known as (i) temporal coherence and (ii) spatial coherence, that correspond to the monochromatic and collimated light output properties. Essentially, optical coherence is the predictability of laser wave properties that stems from the process of stimulated emission [1, 10, 12, 19], due to the generation of photons with "copied" wave traits (phase, wavelength, polarization, and propagating direction) from its spontaneously emitted "seed" photon. Firstly, the **longitudinal optical coherence** corresponds to the predictability of wave traits along the direction of laser wave propagation [1]. Thus, it is a measure of correlation of wave-phases along equidistant wavefronts (red and blue dots as shown in the right of Fig. 1.11) that can indicate whether a light output is **monochromatic** [1]. Monochromatic sources such as lasers have very predictable wave-phases along equidistant wavefronts, where its electric field oscillation component is expected to map out perfect sinusoidal behaviour with propagation time [10]. For this reason, longitudinal optical coherence is also known as **temporal coherence** and can be affected by light scattering, dispersion and diffraction [10]. The widely-accepted standard for a laser's monochromatic width is $\Delta\lambda \leq 1\text{nm}$ [23]. On the other hand, lasers also possess *transverse optical coherence*, which measures the correlation of wave phases extending along a single wavefront [10] (hence, *spatial coherence*). With the knowledge of individual adjacent wavefronts, one can obtain information of the overall beam directionality, as shown on the bottom left side of Fig. 1.11. Ideally, lasers possess a collective of plane waves, which maps out (blue and red arrows) to be perfectly collimated. In spectroscopic measurements, the beam directionality emitted by a gain material can be assessed by performing angular-dependent PL experiments. Lastly, it should be noted that while temporal and spatial coherences both commonly measure wave-phase correlations, they deal with very different aspects. The former traces sinusoidal wave continuity with propagation while the latter traces wavefront distortions. As such, temporal and spatial coherences are ***independent*** of one another.

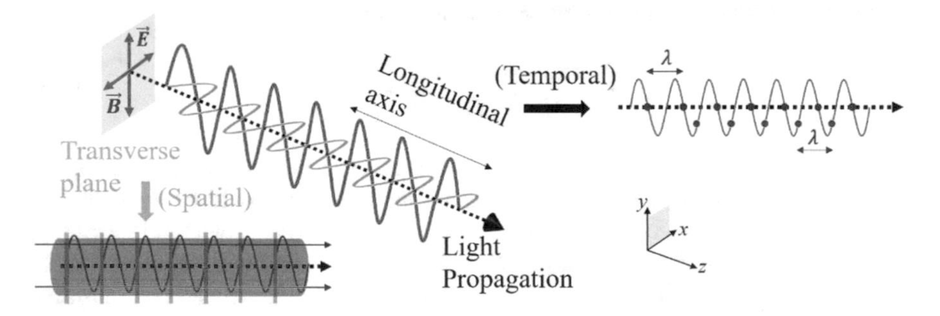

Fig. 1.11 Illustrations of longitudinal (temporal) and transverse (spatial) coherence corresponding to monochromacity and collimation of a laser beam output

Table 1.1 A table of comparison between ASE and Lasing properties. Key similarities and differences between these two phenomena are represented in blue and red, respectively

Properties	ASE	Lasing
Light Amplification (Population Inv)	YES	YES
Optical Feedback	NO	YES
Threshold Behaviour	YES	YES
Output Linewidth	6-12nm	< 1nm
Output Position	Red-shifted from broad Luminescence (e.g. PL)	Lasing modes within vicinity of ASE bandwidth with quick intensity growth
Mechanism	Widened Stimulated Emission	Strict Stimulated Emission
Temporal Coherence (monochromatic)	Strong	Very Strong
Spatial Coherence (Directionality)	Weak/medium	Very Strong

1.4.1 Differentiating Amplified Spontaneous Emission and Lasing

A comprehensive set of protocols was established by Ifor et al. in order to distinguish between lasing output emission and its "look-alikes", for instance, ASE. In recalling from Sects. 1.2 and 1.3, it is clear that the first step to differentiate between them is through the presence (or absence) of a cavity in the measured sample material. Essentially, the lack of waveguiding from optical feedback causes ASE to possess lower degrees of optical coherences as compared to lasing, as presented in Table 1.1. Figure 1.12a illustrates the development of a relatively sharper and more intense ASE spectrum that manifested from unsupported optical gain. Next, Fig. 1.12b illustrates the development of lasing upon provision of optical feedback by externally fabricated cavity, where cavity modes within the gain spectrum evolves to lasing modes with significantly higher intensity than non-participating cavity modes, of which, indicates laser oscillations. In comparison, it is clear that ASE and lasing differ in terms of their (i) spectral widths and (ii) emission features [12, 19, 23]. Nevertheless, the protocols used to identify lasing are as follows [23]: (I) Lasing linewidths of $\leq$1 nm (Temporal Coherence), (II) pump-threshold behaviour followed by superlinear increase in lasing intensity and (III) Visually collimated output (Spatial Coherence), as shown in Fig. 1.12c, d geometries. Two similar artefacts are interference effects in micro-cavities and waveguided light that may also appear spectrally narrow with strong directionality possibly due to substrate leaky-waves [24]. However, since these two artefacts do not occur in the optical gain regime, the trick here is to check

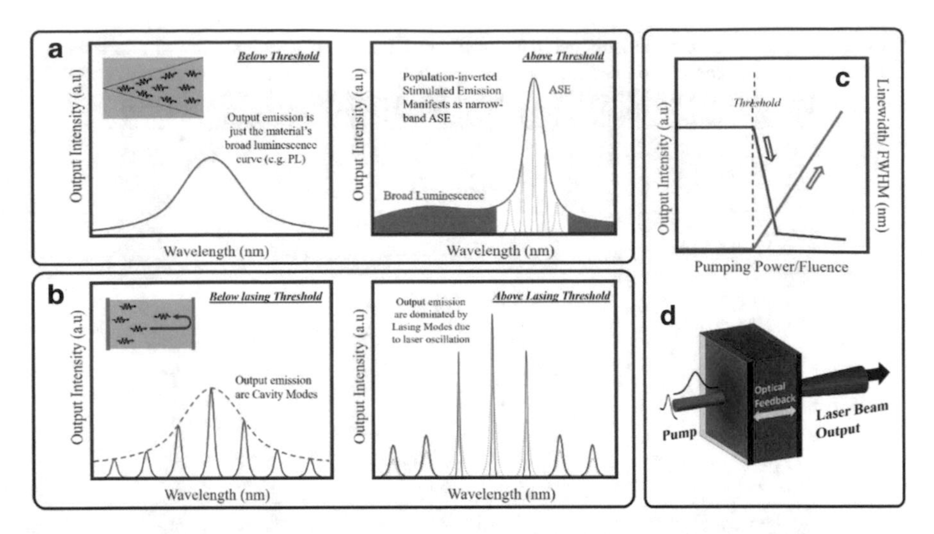

Fig. 1.12 Development of **a** ASE and **b** lasing in the gain medium and cavity system respectively, once pumped above threshold. **c** Linear intensity growth and linewidth narrowing with pump and **d** a visually collimated beam output that indicates lasing output and strong spatial coherence

if a pump-threshold trait exists. Importantly, interference patterns and narrowband spectra are commonly seen in luminescent micro-cavities with edge-waveguiding effects [23]. For microcavity structures, their emission environment can be strongly affected by its self-interference (Etalon effect), causing narrower linewidth emissions and stronger directionality.

1.5 Summary and Conclusions

In this chapter, laser emission is reviewed as a combined effort of population inversion buildup in the pumped gain medium coupled with optical feedback afforded by reflective mirrors in a Fabry–Perot cavity. Ideally, gain media based on the four-level system model have near-zero threshold for optical gain because the lower level does not involve the ground state. However, cavity imperfections such as non-unity mirror reflectivities, reabsorption and etc. manifest themselves as loss channels that compete with the material's intrinsic gain g. Thus, net gain g_{net} leading to laser mode oscillations occurs only when the pumping rate $P > P_{\mathrm{th}}$, leading to the two key steady-state traits shown in Fig. 1.10. Importantly, we also discussed the conceptual differences between ASE and lasing, despite both processes occurring in the optical gain regime. Essentially, the former process leads to a relatively broader spectral width than the latter because of relaxed conditions arising from the lack of optical feedback, resulting in poorer temporal (monochromacity) and spatial (beam divergence) properties.

References

1. R.S. Quimby, *Photonics and Lasers: An Introduction* (Wiley, 2006)
2. S.M. Cases, K. Jianwattananukul, K. Saenyot, S. Lekchaum, K. Locharoenrat, Portable laser 1-D barcode scanner for material identification. Mater. Today Proc. **5**(7), 15143–15148 (2018)
3. L.M. Snell, Monitoring temperatures in concrete construction using IR thermometers. Concr. Int. **37**(1), 57–63 (2015)
4. D. Huber, B. Akinci, P. Tang, A. Adan, B. Okorn, X. Xiong, Using laser scanners for modeling and analysis in architecture, engineering, and construction, in *2010 44th Annual Conference on Information Sciences and Systems (CISS)* (IEEE, 2010), pp. 1–6
5. R.A. Convissar, *Principles and Practice of Laser Dentistry-E-Book* (Elsevier Health Sciences, 2015)
6. H.K. Soong, J.B. Malta, Femtosecond lasers in ophthalmology. Am. J. Ophthalmol. **147**(no. 2), 189–197.e2 (2009)
7. I. Amiri, A.N.Z. Rashed, A.M. Abd Elnaser, E.S. El-Din, P. Yupapin, Spatial continuous wave laser and spatiotemporal VCSEL for high-speed long haul optical wireless communication channels. J. Opt. Commun. **1**(no. ahead-of-print) (2019)
8. K. Johnson, M. Hibbs-Brenner, W. Hogan, M. Dummer, Advances in Red VCSEL technology, in *Advances in Optical Technologies* (2012)
9. D.B. Todorov, Development of hybrid UV VCSEL with organic active material and dielectric DBR mirrors for medical, sensoric and data storage applications. (2009)
10. B.E. Saleh, M.C. Teich, *Fundamentals of Photonics* (Wiley, 2019)
11. Wikiwand, Ruby Laser. https://www.wikiwand.com/en/Ruby_laser. Accessed 29 Mar 2020
12. O. Svelto, D.C. Hanna, *Principles of Lasers* (Springer, 2010)
13. M. Fox, *Optical Properties of Solids* (American Association of Physics Teachers, 2002)
14. L. Casperson, A. Yariv, Spectral narrowing in high-gain lasers. IEEE J. Quantum Electron. **8**(2), 80–85 (1972)
15. K. Shaklee, R. Leheny, Direct determination of optical gain in semiconductor crystals. Appl. Phys. Lett. **18**(11), 475–477 (1971)
16. C. Lange et al., The variable stripe-length method revisited: improved analysis. Appl. Phys. Lett. **91**(no. 19), 191107 (2007)
17. Y. Chan et al., Blue semiconductor nanocrystal laser. Appl. Phys. Lett. 86(no. 7), 073102 (2005)
18. G. Xing et al., Low-temperature solution-processed wavelength-tunable perovskites for lasing. Nat. Mater. **13**(5), 476–480 (2014)
19. A.E. Siegman, *Lasers University Science Books*, vol. 37, no. 208 (Mill Valley, CA, 1986), p. 169
20. O. Svelto, S. Taccheo, C. Svelto, Analysis of amplified spontaneous emission: some corrections to the Linford formula. Opt. Commun. **149**(4–6), 277–282 (1998)
21. L. Feng, Z.J. Wong, R.-M. Ma, Y. Wang, X. Zhang, Single-mode laser by parity-time symmetry breaking. Science **346**(6212), 972–975 (2014)
22. Y. Xiao et al., Single-nanowire single-mode laser. Nano Lett. **11**(3), 1122–1126 (2011)
23. I.D.W. Samuel, E.B. Namdas, G.A. Turnbull, How to recognize lasing. Nat. Photon. 3(no. 10), 546–549 (2009). https://doi.org/10.1038/nphoton.2009.173
24. D. Yokoyama, M. Moriwake, C. Adachi, Spectrally narrow emissions at cutoff wavelength from edges of optically and electrically pumped anisotropic organic films. J. Appl. Phys. **103**(no. 12), 123104 (2008)

Chapter 2
The Halide Perovskite Gain Media

2.1 Introduction to Halide Perovskites

Perovskites rose to fame due to its rapid improvements in solar-cell conversion efficiencies, starting from 3.8% [1] (in 2009) to 22.1% (in 2014) [2] and exceeding 25% (in 2019) [3, 4] while silicon solar cells took several decades of intense research efforts to reach conversion efficiencies ~25% [5]. In 2014, amplified spontaneous emission (ASE) was serendipitously discovered in Perovskite [6], which triggered massive efforts to study its viability to support continuous wave lasing [7]. In this chapter, we begin with a formal introduction to the Perovskite gain material by reviewing its general crystal structure and its structural dimensionality. Importantly, we make a clear distinction between the Perovskite's structural and morphological dimensionality. Next, we provide context to the research field of Perovskite lasing by reviewing the research trajectory and milestones; and comparing them to other classes of gain materials.

2.1.1 The Perovskite Crystal Structure

The term "Perovskite" refers to any semiconducting material with a general chemical formula of $A_{4-m}BX_{6-m}$ [8], where, m is an integer called the Perovskite's structural dimensionality number. By nomenclature, the A, B and X ions are the monovalent cation, metallic divalent cation and monovalent halide anion, respectively [8]. The most common system is the structural three-dimensional Perovskites ($m = 3$), whose crystal structure is shown in terms of the unit cell and octahedral tessellation representation in Fig. 2.1a and b, respectively. In the unit cell representation, the B-cation (orange) resides in the innermost-center of the octahedron cage created by six X anions (blue), forming the octahedral $[BX_6]^{4-}$ building block that is further caged by an "external cube" created by eight corner-sharing A-cations (gray). Spatially, the $[BX_6]^{4-}$ octahedron is situated in the external cube's center, where each X-halide

Y. K. E. Tay et al., *Halide Perovskite Lasers*, Nanoscience and Nanotechnology,
https://doi.org/10.1007/978-981-16-7973-5_2

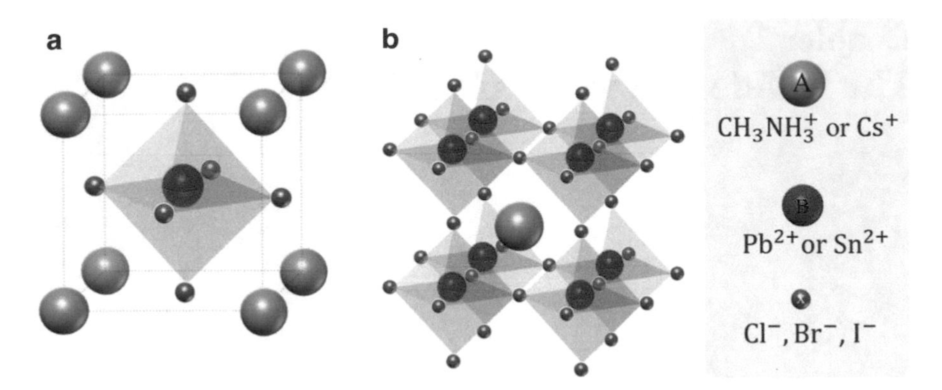

Fig. 2.1 Structural three-dimensional Perovskites of ABX_3 in **a** cubic unit-cell and **b** octahedra-tessellated representations. The $[BX_6]^{4-}$ octahedron is a Perovskite's building block

anion resides at the center of each of the external cube's six surfaces. Alternatively, the Perovskite structure can be presented in the octahedral tessellated representation, where $[BX_6]^{4-}$ octahedra tessellates in all three x, y, z directions, with the A-cation filling the central void. Common choices of A, B and X ions are shown in the grey box of Fig. 2.1 [9]. In this book, we focus our attention on the excellent optical gain properties and its underlying carrier dynamics in lead-halide ($B = Pb^{2+}$) Perovskite systems.

Generally, the overall stability of the Perovskite crystal depends greatly on the mutual suitability of A, B and X ions in terms of their atomic or molecular radii and are characterised by the (i) Goldschmidt's Tolerance Factor t (GTF) and (ii) the Octahedral Tolerance Factor μ (OTF). The GTF and OTF are defined as [10]:

$$t = \frac{r_A + r_x}{\sqrt{2}(r_B + r_x)} \tag{2.1}$$

$$\mu = \frac{r_B}{r_x} \tag{2.2}$$

where r_j is the atomic or molecular radii of the ion j. Here, the GTF calculates the total structural stability while the OTF calculates the $[BX_6]^{4-}$ octahedral stability of a chosen combination of A, B and X ions. Table 2.1 summarises the GTF and OTF values of various Perovskite media [10] and Table 2.2 summarises its corresponding crystallographic phase. For lead halide Perovskites, we fix the B-cation to be Pb^{2+} and X-anions to be Cl^-, Br^- or I^-. Ideally, we want the A-cation to be comfortably larger than Pb^{2+} so that it crystalises in **cubic** phase ($\alpha-$ (pm$\bar{3}$m)) [11], where $0.89 < t_{cubic} < 1$. A classic example of room-temperature cubic-phased Perovskite is the green-emitting MethylAmmonium lead Bromide ($CH_3NH_3PbBr_3$ or $MAPbBr_3$) system [10, 12, 13]. If a chosen A-cation is smaller than Pb^{2+}, then the system crystalises in the **orthorhombic** phase ($\gamma-$ (pnma)) [11], with a GTF range $0.71 < t_{ortho} < 0.89$. This occurs as a result of loosely-fitting A-cation in the central void

Table 2.1 First-principles calculations of the (A) OTF and (B) GTF in common Perovskite assemblies at room-temperature. Data is adapted from [10]

(A)

Cl^-	octahedral factor μ
Ge^{2+}	0.395
Sn^{2+}	0.622
Pb^{2+}	0.535
Ca^{2+}	0.497
Sr^{2+}	0.627
Tm^{2+}	0.503
Sm^{2+}	0.551
Yb^{2+}	0.465
Dy^{2+}	0.578
Br^-	
Ge^{2+}	0.372
Sn^{2+}	0.587
Pb^{2+}	0.500
Ca^{2+}	0.464
Sr^{2+}	0.592
Tm^{2+}	0.474
Sm^{2+}	0.439
Yb^{2+}	0.449
Dy^{2+}	0.515
I^-	
Ge^{2+}	0.350
Sn^{2+}	0.441
Pb^{2+}	0.468
Ca^{2+}	0.418
Sr^{2+}	0.536
Tm^{2+}	0.432
Sm^{2+}	0.505
Yb^{2+}	0.423
Dy^{2+}	0.441

(B)

molecular cation	NH_4^+	HY^+	HA^+	MA^+	FA^+	GUA^+	AZ^+	$DiMA^+$	EA^+	AA^+	$TetraMA^+$	IM^+	$TriMA^+$	$isoPA^+$	PY^+	$isoBuA^+$	$DiEA^+$	PhA^+
ionic radius [Å]	1.70	2.20	2.26	2.38	2.77	2.80	2.84	2.96	2.99	3.00	3.01	3.03	3.04	3.07	3.22	3.60	3.85	3.88
Cl^-																		
Ge^{2+}	0.972	1.110	1.126	1.159	1.266	1.274	1.285	1.318	1.327	1.329	1.332	1.337	1.340	1.348	1.390	1.494	1.562	1.570
Sn^{2+}	0.836	0.955	0.969	0.997	1.089	1.096	1.105	1.134	1.141	1.143	1.146	1.150	1.153	1.160	1.195	1.285	1.344	1.351
Pb^{2+}	0.883	1.008	1.023	1.053	1.150	1.158	1.168	1.198	1.205	1.208	1.210	1.215	1.218	1.225	1.262	1.357	1.419	1.427
Ca^{2+}	0.905	1.034	1.049	1.080	1.179	1.187	1.197	1.228	1.236	1.238	1.241	1.246	1.248	1.256	1.294	1.391	1.455	1.463
Sr^{2+}	0.833	0.951	0.966	0.994	1.085	1.092	1.102	1.130	1.137	1.139	1.142	1.146	1.149	1.156	1.191	1.280	1.339	1.346
Tm^{2+}	0.902	1.030	1.045	1.076	1.175	1.183	1.193	1.223	1.231	1.234	1.236	1.241	1.244	1.251	1.290	1.386	1.450	1.457
Sm^{2+}	0.873	0.998	1.013	1.042	1.138	1.146	1.156	1.185	1.192	1.195	1.197	1.202	1.205	1.212	1.249	1.343	1.404	1.412
Yb^{2+}	0.925	1.057	1.072	1.104	1.205	1.213	1.224	1.255	1.263	1.265	1.268	1.273	1.276	1.284	1.323	1.422	1.487	1.495
Dy^{2+}	0.858	0.981	0.995	1.024	1.119	1.126	1.136	1.165	1.172	1.174	1.177	1.182	1.184	1.191	1.228	1.320	1.380	1.388
Br^-																		
Ge^{2+}	0.961	1.094	1.109	1.141	1.243	1.251	1.262	1.293	1.301	1.304	1.306	1.312	1.314	1.322	1.362	1.462	1.527	1.535
Sn^{2+}	0.831	0.946	0.959	0.987	1.075	1.082	1.091	1.119	1.125	1.128	1.130	1.135	1.137	1.144	1.178	1.264	1.321	1.328
Pb^{2+}	0.879	1.001	1.015	1.044	1.138	1.145	1.154	1.183	1.191	1.193	1.195	1.200	1.203	1.210	1.246	1.337	1.397	1.405
Ca^{2+}	0.901	1.025	1.040	1.069	1.165	1.173	1.183	1.212	1.220	1.222	1.225	1.229	1.232	1.239	1.276	1.370	1.431	1.439
Sr^{2+}	0.828	0.943	0.956	0.984	1.072	1.079	1.088	1.115	1.122	1.124	1.126	1.131	1.133	1.140	1.174	1.260	1.317	1.324
Tm^{2+}	0.894	1.018	1.033	1.062	1.157	1.165	1.174	1.204	1.211	1.214	1.216	1.221	1.223	1.231	1.267	1.360	1.422	1.429
Sm^{2+}	0.916	1.043	1.058	1.088	1.186	1.194	1.204	1.234	1.241	1.244	1.246	1.251	1.254	1.261	1.299	1.394	1.457	1.464
Yb^{2+}	0.910	1.036	1.051	1.081	1.178	1.185	1.195	1.225	1.232	1.235	1.237	1.242	1.245	1.252	1.290	1.384	1.447	1.454
Dy^{2+}	0.870	0.990	1.005	1.033	1.126	1.133	1.143	1.171	1.179	1.181	1.183	1.188	1.190	1.198	1.233	1.324	1.383	1.390
I^-																		
Ge^{2+}	0.927	1.048	1.062	1.090	1.183	1.190	1.200	1.229	1.236	1.238	1.240	1.245	1.248	1.255	1.290	1.381	1.440	1.448
Sn^{2+}	0.869	0.981	0.995	1.022	1.109	1.115	1.124	1.151	1.158	1.160	1.162	1.167	1.169	1.176	1.209	1.294	1.350	1.356
Pb^{2+}	0.853	0.963	0.976	1.003	1.088	1.095	1.103	1.130	1.136	1.138	1.141	1.145	1.147	1.154	1.187	1.270	1.324	1.331
Ca^{2+}	0.883	0.997	1.011	1.038	1.126	1.133	1.142	1.169	1.176	1.179	1.181	1.185	1.188	1.194	1.228	1.314	1.371	1.378
Sr^{2+}	0.815	0.920	0.933	0.958	1.040	1.046	1.054	1.079	1.086	1.088	1.090	1.094	1.096	1.103	1.134	1.213	1.266	1.272
Tm^{2+}	0.874	0.988	1.001	1.028	1.116	1.122	1.131	1.158	1.165	1.167	1.170	1.174	1.176	1.183	1.217	1.302	1.358	1.365
Sm^{2+}	0.832	0.940	0.953	0.978	1.062	1.068	1.077	1.102	1.109	1.111	1.113	1.117	1.119	1.126	1.158	1.239	1.292	1.299
Yb^{2+}	0.880	0.994	1.008	1.035	1.123	1.130	1.139	1.166	1.172	1.175	1.177	1.182	1.184	1.191	1.224	1.310	1.367	1.374
Dy^{2+}	0.869	0.981	0.995	1.022	1.109	1.115	1.124	1.151	1.158	1.160	1.162	1.167	1.169	1.176	1.209	1.294	1.350	1.356

Table 2.2 A summarised table of common Perovskite crystal structures in room temperature conditions. Knowledge of the crystallinity of the Perovskite of interest is essential for verifying its corresponding X-Ray Diffraction angular peaks

Perovskite Crystallography	Orthorhombic $a \neq b \neq c$	Cubic $a = b = c$	Tetragonal $a = b \neq c$
Space Group	$Pbnm$	$Pm\bar{3}m$	$P4/mbm$
Unit Cell Representation			
Octahedral Network Representation (Top-View)			
Goldschmidt's Tolerance Factor range (GTF)	$0.71 < t_{ortho} < 0.89$	$0.89 < t_{cubic} < 1$	$t_{tetra} > 1$
Ionic Compatibility	A-cation too small to fit into B-cations' interstices	A-cation is comfortably larger than the B-cations (Ideal compatibility)	A-cation too big OR B-cation too small
Examples of Perovskite Material	$CsPbBr_3$ [Ref 13,14]	$CH_3NH_3PbBr_3$ [Ref 9-11]	$CsSnBr_3$ [Ref 9]

and an example of room-temperature orthorhombic Perovskite is the Cesium Lead Bromide system ($CsPbBr_3$) [14, 15]. On the other hand, if the chosen A-cation is exceedingly large, then the system crystalises in the mildly-distorted tetragonal phase ($\beta-(p4/mbm)$) [11], with a GTF range of $t_{tetra} > 1$. They are generally unstable and may collapse into the two-dimensional planar (multiple QW-like) structures, such as the Ruddlesden-Popper (RPP), Aurivillius and Dion-Jacobson (DJ) phases [16]. An example of room temperature occurring tetragonal Perovskite is the formamidinium lead bromide system ($(NH_2)_2CHPbBr_3$ or $FAPbBr_3$) [10]. Equally important, the value of an octahedron's OTF indicates the suitability of a B-X ions. The acceptable range of stable $[BX_6]^{4-}$ formation occurs between $0.41 < \mu < 0.90$. Similarly, OTF of commonly chosen B and X ions are summarised in Table 2.1(B).

Intrinsically, the Perovskite's crystal phase adapts with temperature, which leads us to the concept of phase-transition. For example, the cubic $CH_3NH_3PbBr_3$ system transits into orthorhombic phase at below 150 K [12, 13] while the orthorhombic $CsPbBr_3$ transits into tetragonal phase at 361 K and quickly to cubic phase at 403 K [15]. Interestingly, the use of co-existing mixed-phases at phase-transiting temperatures served as a "trick" for replicating a conducive energy-funnelling environment for achieving low-temperature assisted CW pumped lasing [17, 18]. Another way to characterise Perovskite crystallography is the X-ray diffraction (XRD), where Fig. 2.2 shows the XRD spectrum of $CH_3NH_3PbBr_3$ in various morphologies. Due to the limited scope of this book, interested readers are highly encouraged to consult references for XRD peak angular position assignments for the Perovskite material of interest [12, 13, 19–21].

2.1.2 Perovskite Structural Tuning

Perovskite's structural dimensionality is given by the number of axes which the $[BX_6]^{4-}$ octahedra tessellates in, as shown in Fig. 2.3. From the general chemical formula of $A_{4-m}BX_{6-m}$, the three, two, one and zeroth structural dimension Perovskites have chemical formulas of ABX_3, A_2BX_4, A_3BX_5 and A_4BX_6, respectively [27]. Structural **two-dimensional** Perovskites (A_2BX_4, Fig. 2.3b) are formed when the $[BX_6]^{4-}$ octahedra connects in two spatial directions, thereby producing multiple planar layers that are separated by organic ligand spacers via weak van-der waals forces [8]. From an energetic viewpoint, the ligands act as the interlayer's periodic potential barrier [28] and that the collective system resembles a multi-layered quantum well. Structural **one-dimensional** Perovskites (A_3BX_5, Fig. 2.3c) are formed when the $[BX_6]^{4-}$ octahedra connects only in one spatial direction, that resembles multi-rod arrays. Lastly, structural **zero-dimensional** Perovskites (A_4BX_6, Fig. 2.3d) are formed when all $[BX_6]^{4-}$ octahedra are disconnected and isolated. Instead, each $[BX_6]^{4-}$ is charge-neutralised by four neighbouring A-cations.

Another sub-class of Perovskites known as the *Ruddlesden-Popper Perovskites* (RPPs) are also of interest for lasing applications. The RPPs are defined with general structural formula of $(R-NH_3)_2A_{n-1}B_nX_{3n+1}$ [29], where R and n are the organic

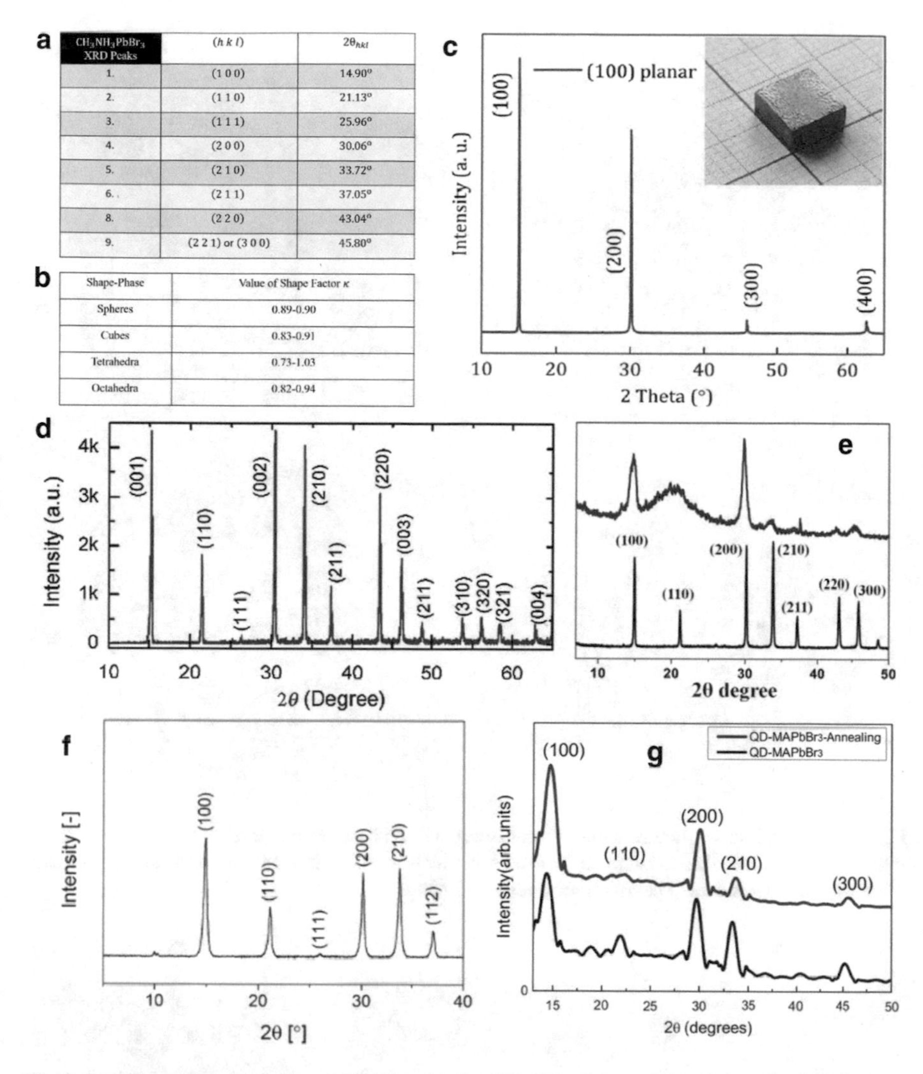

CH₃NH₃PbBr₃ XRD Peaks	(h k l)	$2\theta_{hkl}$
1.	(1 0 0)	14.90°
2.	(1 1 0)	21.13°
3.	(1 1 1)	25.96°
4.	(2 0 0)	30.06°
5.	(2 1 0)	33.72°
6.	(2 1 1)	37.05°
8.	(2 2 0)	43.04°
9.	(2 2 1) or (3 0 0)	45.80°

Shape-Phase	Value of Shape Factor κ
Spheres	0.89-0.90
Cubes	0.83-0.91
Tetrahedra	0.73-1.03
Octahedra	0.82-0.94

Fig. 2.2 **a** Example of calculated XRD peaks for $CH_3NH_3PbBr_3$ nanocrystals and **b** the corresponding shape-factor values in the Scherrer's equation, based on the observed physical nanocrystallites from transmission electron microscopy (TEM). XRD patterns of **c** bulk single-crystal [22], **d** powder [23], **e** spin-coated thin-film (Black) [24], **f** nanocrystals [25] and **g** quantum dots [26] of the $CH_3NH_3PbBr_3$ system. Due to the pristine quality of single crystals, only diffraction peaks along (h 0 0) axes are well-defined

alkyl group and the tessellated thickness of each planar layer, respectively, as shown in Fig. 2.4. Thus, n = 1 corresponds to a pure two-dimensional Perovskite structure while n = 2 corresponds to a "modified" two-dimensional Perovskite structure, where each octahedral plane is stacked twice before successive organic spacers. As n → ∞, it simply corresponds back to the three-dimensional structure. For this reason, RPPs

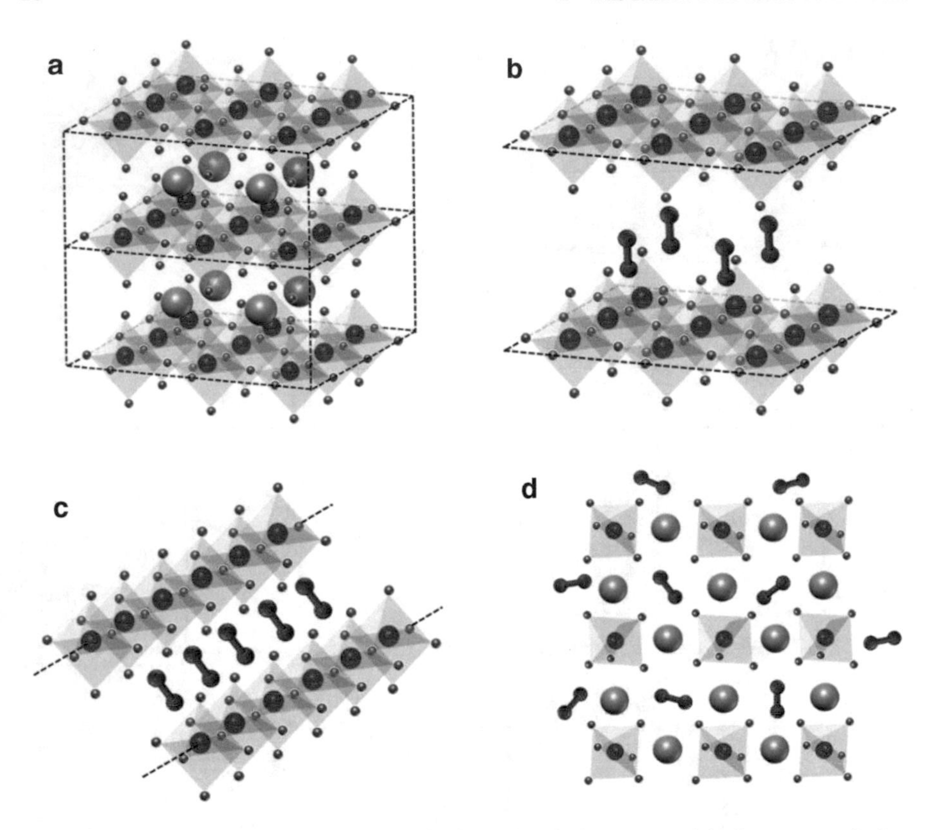

Fig. 2.3 Perovskite's structural dimensionality depends on its interconnectivity between octahedra-networks. **a** Three-dimensional, **b** two-dimensional, **c** one-dimensional and lastly, **d** zero-dimensional Perovskites. The green icons denote ligands

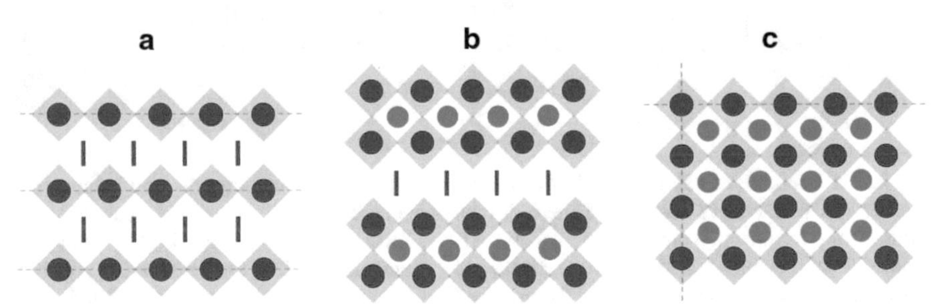

Fig. 2.4 The Ruddlesden-Popper Perovskite (RPP) structure. **a** $n = 1$ (pure two-dimensional), **b** $n = 2$ and **c** $n = \infty$ (pure three-dimensional). Between successive planes, periodic potential barriers are set up by the ligand spacers (green rectangles)

are otherwise known as quasi two-dimensional or "layered" Perovskites in literatures. Later in Sects. 2.2.1 and 2.2.3, we will see that RPPs are excellent gain media due to their enhanced ambient stability and excitonic binding energies [27, 30–32], high carrier mobilities for devices [1, 33–35] and suppressed ion migration [36]. Others deem them unsuited for lasing due to efficient charge transfer at the organic interfaces that competes with its ASE buildup [37].

2.1.3 Research Interests in Perovskite Lasers

It is important for readers entering the field of Perovskite lasing to have an overview of the current research interests, milestones achieved and challenges that remains to be tackled. Generally, an excellent gain material is one with access to emission wavelength tunability, low pumping threshold for lasing, and single mode operation. In comparison to organic (polymers and insoluble small molecules) and inorganic (chalcogenide quantum dots (CQDs) and group III-nitrides) semiconductors, Perovskites have consistently shown to be satisfying these qualities [38]. For instance, although vacuum vapor-deposition commonly employed to fabricate organic emitters (OLEDs) based on small molecules (such as $Al(C_9H_6N)_3$) [39] allow for ease of fabricating complicated multilayers with sustained device performance [40, 41], they generally suffer from high costs of production [42] and limited scalability due to pixilation using evaporation masks. In contrast, organic polymers offer the advantages of low cost of production via solution processing and compatibility with inkjet printing for large-area device scaling, but they generally suffer from relatively lower device efficiencies [40]. In recent years, although scientists have increased efforts to develop solution processable small molecule based devices that are comparable to its vapor deposited counterpart, its success is limited by lower material packing density and shorter carrier lifetimes than the latter [40]. On the other hand, inorganic semiconductors generally face serious problems with emission wavelength tunability. For instance, chalcogenide quantum dots (CQDs) such as CdSe and ZnS have difficulty acquiring blue lasing, due to its small dot-size. Essentially, its strong quantum confinement makes auger recombination rates highly competitive and its high volume-to-surface ratio makes it susceptible to high surface defect densities [43]. In addition, their surface insulating ligands cause severe charge transport issues [44]. On the other hand, group III-nitrides such as the Ga-rich $In_{1-x}Ga_xN$ typically suffer from the so-called "green-valley of death" [45, 46] or "green gap" [47], which describes the rapid decrease in external quantum efficiency (EQE) [48] when the emission shift towards blue to green as molar ratio of In increases >0.3 [49]. Essentially, carrier leakage and Crystalline defects arising from growth temperature, stacking faults and lattice mismatch between GaN and InGaN are problems that remain to be tackled, before the implementation of stable group III-nitride based green laser diodes [47].

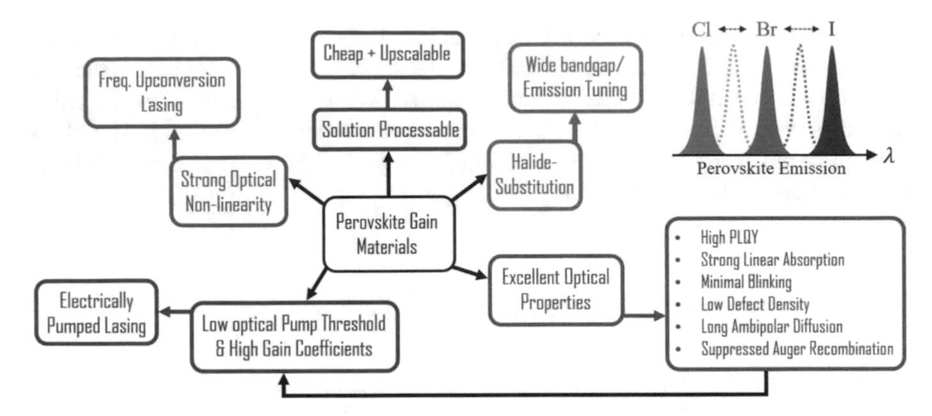

Fig. 2.5 Overview of the attractive properties of Perovskites as gain media

Figure 2.5 presents an overview of the attractive properties of Perovskites, making them a family of extremely promising candidates for sustainable low-energy consumption and wavelength tunable lasers. Firstly, the access to **low-temperature solution-processed synthetic routes** and its up scalability for industrial production [50] grants the advantage of lower production costs that surpasses organic small molecule emitters. Secondly, the ease of halide substitution [6, 51, 52] pre- or post-synthesis allow for a wide visible range tunability in light emission, that could easily overcome the "green valley of death" in group III-nitrides and the strong quantum confinement induced problems of blue emission in CQDs. Thirdly, it possesses a range of excellent optical properties [53], ranging from high photoluminescence quantum yield (PLQY) to minimal blinking [54] and to long ambipolar carrier diffusion lengths, which make them extremely suitable for lasers and optoelectronic applications. Fourthly, their low pumping threshold [55], high g_{net} [25, 56] and strong optical non-linearity [25, 57] make them extremely promising in frequency-upconverted lasers that can find potential applications in biomedical photonics [58] without having to meet cumbersome phase-matching conditions [57, 59].

The holy grail in this field is to realise **electrically-driven Perovskite lasers**, where material optimisation and careful device-stack planning are extremely important. Towards this goal, early Perovskite lasing studies began with optical pumping using femtosecond laser pulses, which would later see a progress to longer duration pulses such as nanosecond to even continuous wave (CW) excitations [53, 60–64]. The common belief is that the establishment of stable CW pumped Perovskite lasing is needed as a steppingstone before heading over to electrically pumped Perovskite lasing. However, as shown in the top wing of Fig. 2.6, the longer excitation pulse duration introduces two problems: (i) lower peak excitation intensity that causes increased pump threshold for lasing [60] and (ii) increased heating effects that exacerbates due to the Perovskite's poor thermal conductivity [7, 53]. In addition, the bottom wing of Fig. 2.6 shows an overview of other related research problems being

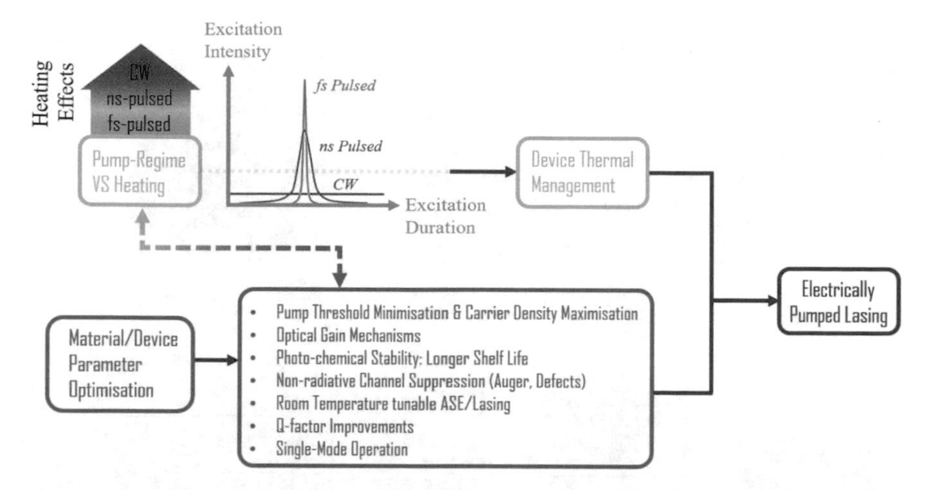

Fig. 2.6 An overview showing the coupled effort (gray-green arrow) of device thermal management and optimisation strategies will propel the field towards electrically pumped Perovskite lasers. The top arrow (blue-cool to orange-hot) indicates increased thermal damage on Perovskite layers with elongated pump durations

tackled since 2014 till present, which must be overcome before the realisation of electrically pumped Perovskite laser.

In the earlier years, the use of femtosecond pulsed excitation was crucial in understanding and resolving the interplay between sub-picosecond processes such as auger recombination [25, 65–68], carrier thermalisation [69, 70], photon recycling [71, 72] and optical build-up processes [73–78]. This is because a deep understanding of the interplaying processes promoting and/or hindering optical gain can provide us with insights essential for developing optimisation strategies and engineering a working device structure [60]. On the other hand, nanosecond and longer duration pulses are termed "*Quasi-CW*" pumping as the above carrier dynamics would have been inhibited by the relatively longer pump event. As early as 2015, Cadelano and co-workers [7] determined that $CH_3NH_3PbI_3$ thin films would require ~50 kW cm^{-2} of intense 300 ns pulsed excitation while maintained under 180 K conditions, so as to mitigate any thermal degradation of its active layer. Importantly, the work highlights an additional need for incorporating efficient thermal management strategies [7, 53].

Figure 2.7 shows a statistical overview of lasing reports in three-dimensional Perovskite media from 2014 to 2020. Figure 2.7a shows a rapidly growing amount of Perovskite lasing reports with time, showing its progressive nature of the field. Certainly, its fame can be attributed to the influence of other studies, ranging from solar cells [3, 79–82], photodetectors [83–87], field-effect transistors (FETs) [88, 89] and LEDs [90–94]. Figure 2.7b shows the Perovskite morphological system breakdown, which reveals two dominant domains of interest: (i) polycrystalline thin films and (ii) nanocrystals. Although small in proportion, but the nanoimprint lithographic and inkjet-printing techniques are highlights in recent years, which saw the success of a prototypical single-mode Distributed FeedBack (DFB) laser. Importantly, these

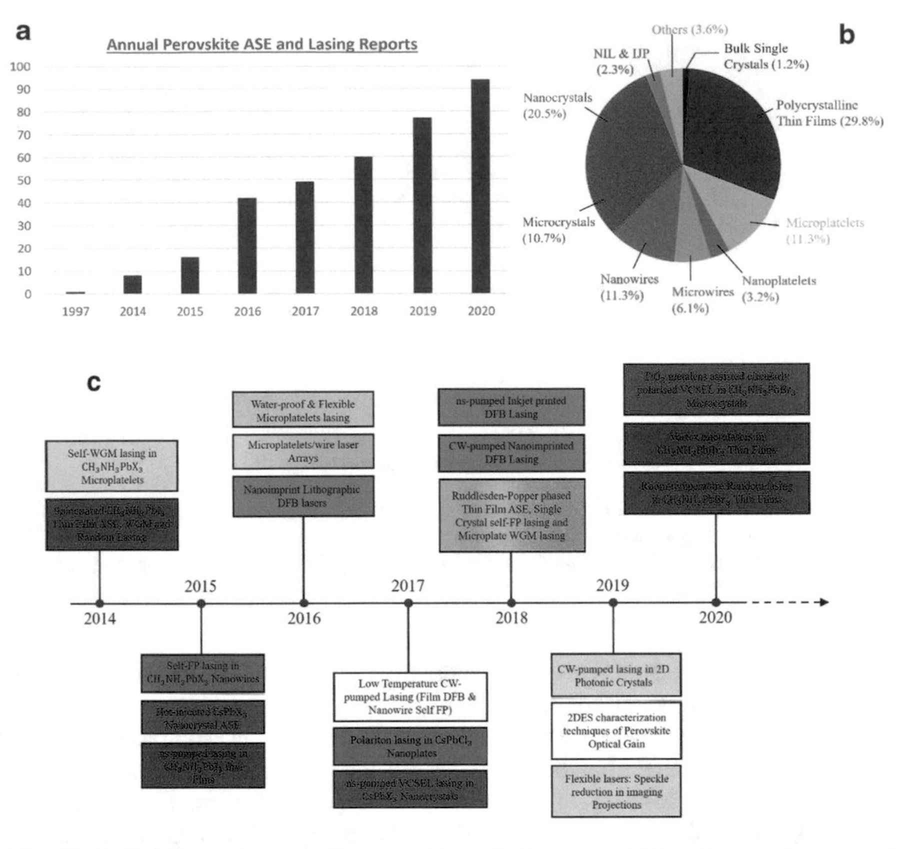

Fig. 2.7 Statistical overview and milestones of the optically pumped ASE and lasing. **a** The upward trend in annual published reports, **b** the distribution of ASE and lasing reports in various Perovskite morphologies and **c** a milestone timeline comprising of research highlights from 1997 till 2020

techniques, albeit new, hold potential in the upscaling and production of large-area Perovskite lasers in the future. Perovskite micro-morphologies are also interesting lasing systems, in which their geometrical structure gives rise to waveguiding optical feedback. Figure 2.7c illustrates a timeline of the important milestones. Generally, the field displays an early interest in establishing and optimising synthetic routes for a myriad of Perovskite morphologies, together with a concurrent shift into using longer excitations, such as from fs-pulses to ns-pulses. In 2017, the field ventures into CW pumping, which highlighted the lasing death phenomenon and also into exciting areas of polariton lasing with a promised lower lasing threshold arising from the so-called "polariton condensate" than conventional photon lasing arising from population inversion mechanisms. In Chap. 3, we shall see that a strong morphological anisotropy and high transition oscillator strength are demanded for the successful demonstration of polariton lasing. From 2018 onwards, the field progressed with new-found interests in upscaling the production of Perovskite based laser devices, with the introduction of inkjet-printing and nanoimprinted lithography. Although not shown, the field in current status, is also in development stages of studying CW pumped Perovskite lasing at room temperature, without the reliance on cryogens.

2.2 Optical Gain in Perovskite Morphologies

In this section, we shall discuss the observation of optical gain in each of the morphological categories shown in Fig. 2.7b. We will also introduce several interesting alternative synthetic routes such as nano-imprint lithographic (NIL) [95–97] and inkjet-printing (IJP) [98–100] techniques, although their advantages in device fabrication for Perovskite lasers will be discussed in Chap. 3. Importantly, we highlight the importance of synthesis optimisation intended to acquire higher PLQY [25], improved surface passivation [75] and/or homogeneity [101] that directly aids in reducing pumping thresholds [75] of ASE and lasing.

2.2.1 Perovskite Single Crystals

Perovskite single crystals (PSCs) generally refer to its bulk-grown monocrystalline morphology that offer much lower trap-densities at $\sim 10^9 - 10^{11}$ cm^{-3} [102, 103], micron-ranged carrier diffusion lengths [102, 104] and slower non-radiative recombination rates [55]. Thus, these qualities make PSCs an attractive testbed for optical gain studies. However, in a report by Wu et al., it was determined that the surface and bulk-volume carrier dynamics are vastly different, with the former behaving closer to polycrystalline nature [105] while the bulk-volume suffers from strong self-absorption effects. Furthermore, its large physical dimension (~several mm^2) would reduce the resulting device's compactness [53], all of which would complicate the success of PSC lasers. Yet, multi-photon pumped ASE [106, 107] and subsequently ultralow temperature random lasing [108] and room-temperature lasing [30] have been observed in these PSC systems. In 2016, Yang et al. was the first to report ASE in $CH_3NH_3PbI_3$ PSCs under two- (2PA) and three-photon absorption (3PA) regimes under 77 K conditions [106], as shown in Fig. 2.8a–c, respectively, with associated thresholds in orders of mJ cm^{-2}. Successful frequency-upconverted ASE was attributed to a giant 3PA non-linear coefficients at $\gamma = 4.76 \times 10^{-4}$ cm^3 GW^{-2} [106]. Importantly, cryogenic conditions were applied to induce crystal phase-transition in the $CH_3NH_3PbI_3$ PSC, so as to reduce auger recombination rates originating from $CH_3NH_3^+$ cation disorder and spin–orbit coupling [106]. Soon after in 2019, Zhao et al. was the first to also report multiphoton pumped ASE in carefully grown $CsPbBr_3$ PSC-rods, with an unsaturated gain determined at $g \sim 38$ cm^{-1} and a low 2PA ASE pumping threshold at 0.65 mJ cm^{-2} [107], as shown in Fig. 2.8d–f. Remarkably, the $CsPbBr_3$ PSC-rods demonstrated stable ASE emissions for up to nearly 40 h, thereby highlighting its robust potential [107]. A quick glance at Fig. 2.8 shows that ASE in both hybrid and fully inorganic PSCs seems to be relatively blue shifted from the

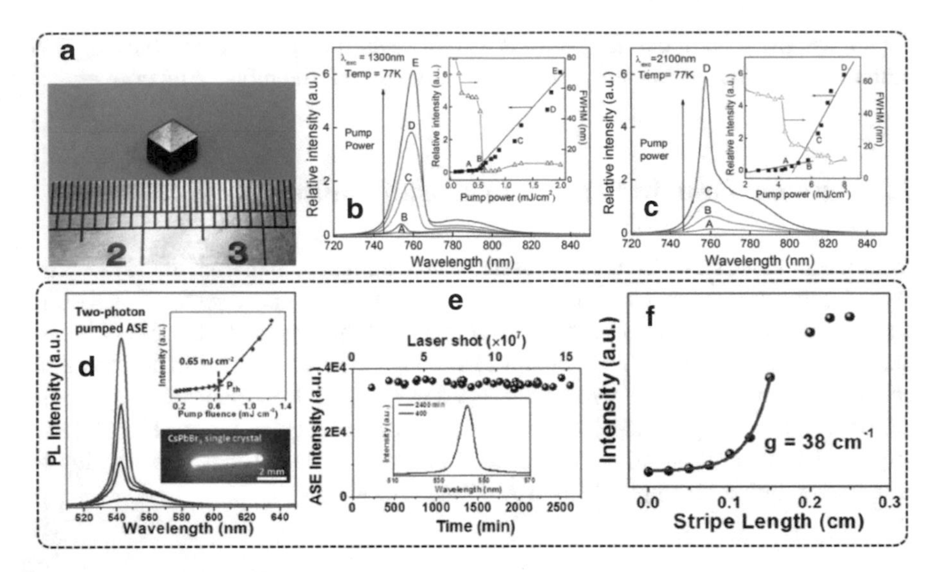

Fig. 2.8 Multi-photon pumped ASE in **a–c** $CH_3NH_3PbI_3$ [106] and **d–f** $CsPbBr_3$ PSCs [107]. **a** The close-up view of a $CH_3NH_3PbI_3$ PSC, with its **b** two- and **c** three-photon pumped ASE spectra at 77 K, respectively. **d** The two-photon pumped ASE spectra of a $CsPbBr_3$ PSC-rod, with the inset showing its pumping threshold and an image of the rod in light amplification regime. **e** The highly stable ASE emission kept up to 40 h, with a corresponding measured unsaturated gain coefficient of $g \sim 38\,cm^{-1}$

broad PL spectral position. However, Zhao et al. argued that this is only an apparent visual judgement caused by an already red-shifted broad PL due to strong reabsorption effects [107]. Instead, Zhao et al. proposed that the observed ASE most likely originates from the conventional biexcitonic gain mechanisms [107], which will be discussed in Chap. 3.

On the other hand, lasing in RPP bulk single crystals (RP-PSCs) of $(BA)_2(MA)_{n-1}Pb_nI_{3n+1}$ was first reported in 2018, by Chinnambedu et al. As shown in Fig. 2.9a–c, growing RP-PSCs with increasing planar-stacks n results in tunable red-shifted lasing emission. Importantly, as shown in Fig. 2.9b, the lasing results from a natural wave-guiding effect arising from the "step-like" pyramidal contours along the crystal surface, with edge-bending guided by total internal reflection. Thus, Chinnambedu et al. assigned the observed lasing emissions in Fig. 2.9c to whispering gallery mode (WGM) lasing [30]. Although relatively new to the field, RPP systems are excellent candidates for lasing applications, due to their (i) enhanced ambient stability [27, 31] and (ii) stronger excitonic properties [27] afforded by the dielectric contrasts between the organic ligands and successive planar stacks [27]. Generally, strongly excitonic emitters can benefit from a consistent PLQY that is independent on photogenerated carrier densities [61]. Strongly directional random lasing (RL) shown in Fig. 2.9d–f was also observed in surface-blemished $CH_3NH_3PbBr_3$ PSCs at 4 K cryogenic conditions. where the surface cracks and imperfections behaved as random resonator sites for optical confinement [108].

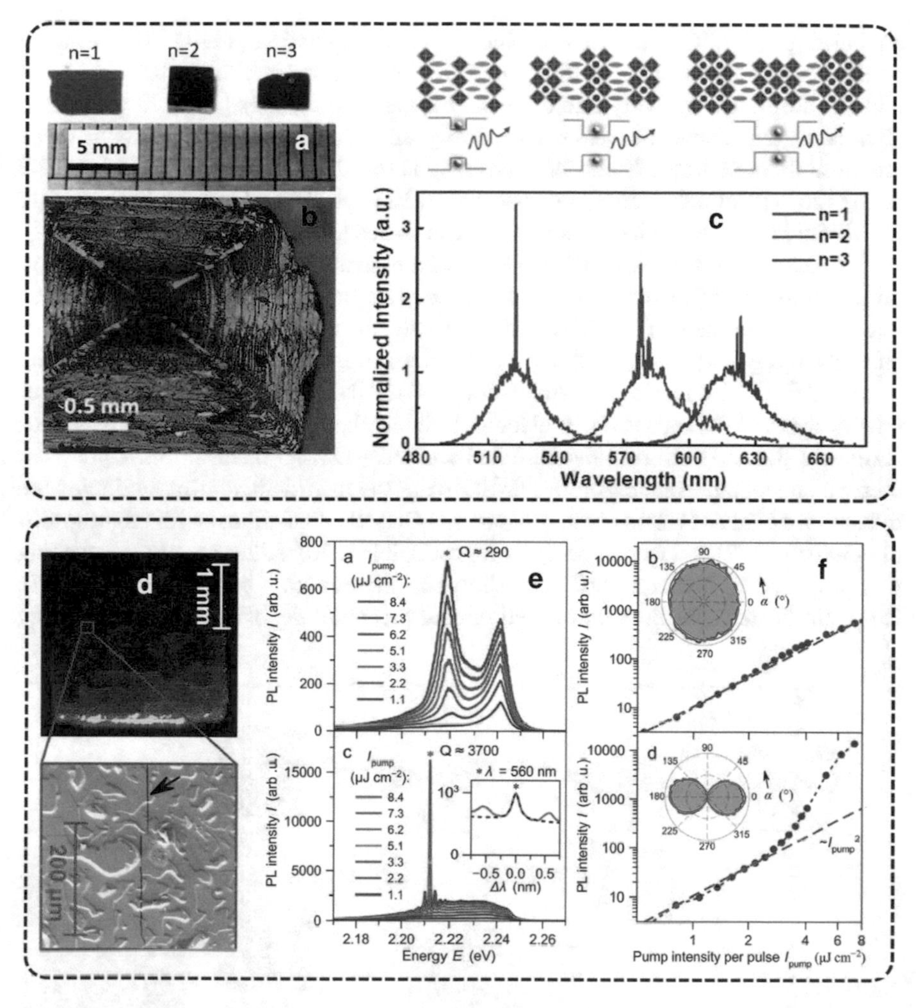

Fig. 2.9 **a–c** As-grown image of n = 1–3, optical image of n = 1 and tunable WGM lasing in $(BA)_2(MA)_{n-1}Pb_nI_{3n+1}$ RP-PSCs [30]. **d–f** Random lasing in $CH_3NH_3PbBr_3$ PSCs at 4 K conditions, as a result of surface blemishes acting as optical confinement sites, revealing excellent Q ~ 3700 with strong lasing directionality once pumped above threshold [108]

2.2.2 *Perovskite Thin-Films*

Based on Fig. 2.7b, lasing from Perovskite Thin Films (PTFs) account for nearly a-third of total published reports, partly due to their fame in solar cells. Unlike PSCs, PTFs are polycrystalline in nature, where "monocrystalline islands" called grains are separated by grain boundaries (pinholes) [109]. As seen in the top wing (black), PTFs conventionally synthesized in the early days are directly spincoated (under inert nitrogen environments [110]) from dissolved precursors and subsequently thermal annealed at ~100 °C to ensure complete solvent evaporation [111–113]. However, studies showed that the films suffer from (i) incomplete film coverage [101, 114, 115] and (ii) film inhomogeneity due to abundant grain boundaries [109]. In particular, abundant grain boundaries causes generally smaller grains and are shown to be responsible for undesirable carrier quenching [116, 117], which greatly hinders PTF lasing [20, 118]. Furthermore, the potential of PTFs is plagued by ambient moisture instability [31, 119] and halide ion migration issues [36, 120, 121].

Eventually, two modifications: (1) anti-solvent treatment and (2) hot-casting were proposed to maximise grain sizes and reduce grain boundaries [114]. In 2014, Jeon et al. [122] introduced the addition of anti-solvents such as toluene [110, 122] or chlorobenzene [74] during mid-stage of spincoating, as shown in the middle wing (blue) of Fig. 2.10. During the spincoating of $CH_3NH_3PBI_3$ PTFs, the addition of toluene causes MAI and DMSO molecules to intercalate between the PbI_2 layers, thus forming a flat and uniform intermediate MAI-PbI_2-DMSO phased film. Upon post thermal annealing, evaporation of DMSO from the intermediate film facilitates the formation of flat and relatively homogeneous $CH_3NH_3PBI_3$ films, with larger grains of ~500 nm [122] that were reportedly reproducible [110]. Alternatively, hot-casting was proposed by Nie et al. [101], as shown in the bottom wing (red) of Fig. 2.10. Here, pre-heated substrates were used during precursor deposition and spincoating

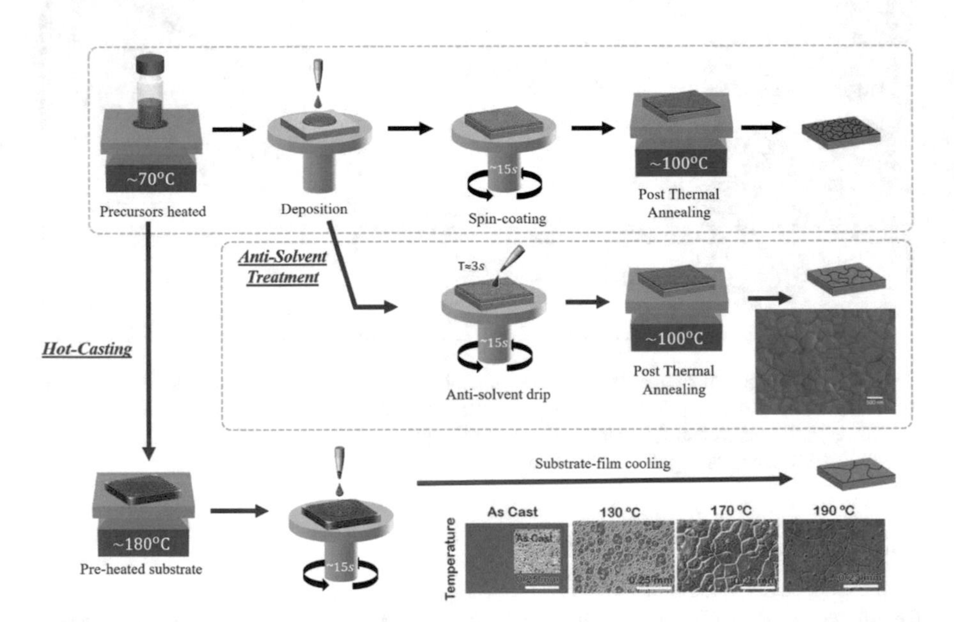

Fig. 2.10 An overview of spincoated PTFs. Black (Conventional): direct spincoating of AX and PbX_2 dissolved precursors onto a clean substrate followed by post thermal annealing up to 1 h. Blue (Anti-solvent treatment [122]): modified method where non-polar antisolvents such as toluene is dripped midway during spincoating. Red (Hot-casting [101]): another modified method in which the substrate is pre-heated on a hotplate at ~180 °C and dissolved precursors are swiftly spincoated and left to cool

to allow solvent retainment that permits prolonged self-assembly of Perovskite crystallisation during spincoating [101]. Giant ~1–2 mm grains of $CH_3NH_3PbCl_xI_{3-x}$ PTFs were reported via the hot-casting pathway [101]. Subsequently, a study of correlation between PTF morphology and optical gain constants were conducted by Lafalce et al., who uncovered that chloroform-treated $CH_3NH_3PbBr_3$ PTFs possessed twice as much unsaturated gain coefficient ($\sim$315 cm^{-1}) and lower lasing ASE pumping thresholds than its untreated counterpart [123]. He proposed that this is due to the synergistic combination of (i) reduced pump scattering (optically smooth), (ii) overall higher PLQY and (iii) reduced waveguiding losses in the chloroform-treated $CH_3NH_3PbBr_3$ PTFs [123]. In Chap. 3, we shall introduce nanoimprint lithographic [96] and inkjet printing [98] approaches for producing well-controlled and optically smooth PTFs required for fabricating low threshold Perovskite ASE and lasing [96, 98, 124].

In 2014, Xing et al. was the first to show remarkably stable ASE (~788 nm) up to ~24 h in conventionally spincoated $CH_3NH_3PbI_3$ PTFs, excited using 150 fs 600 nm pulses [6]. As shown in Fig. 2.11, an ultralow ASE pumping threshold of

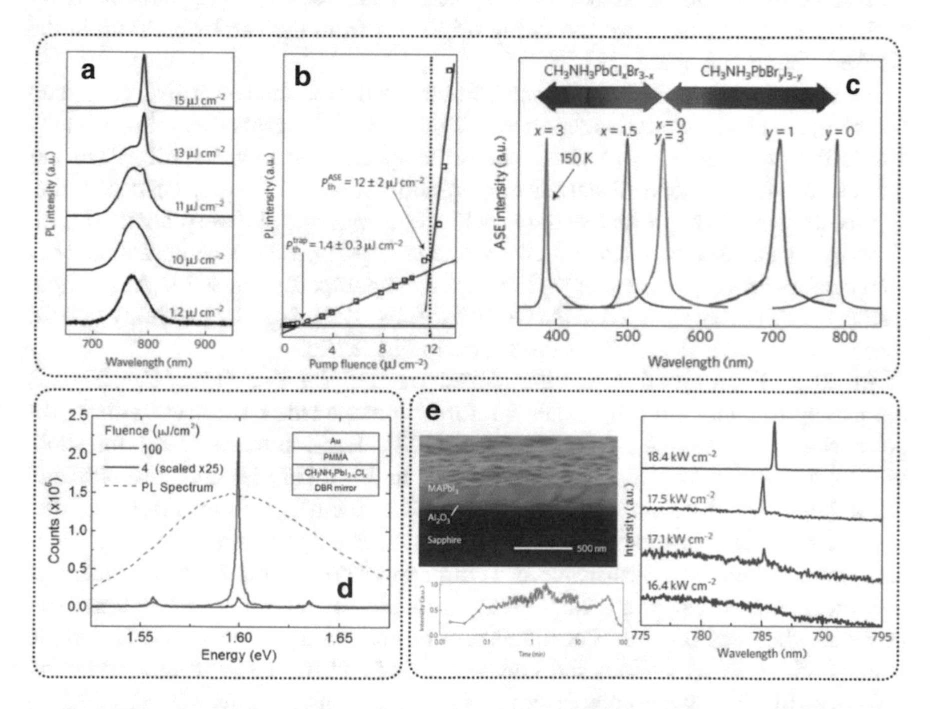

Fig. 2.11 Early reports of optically pumped ASE and lasing in $CH_3NH_3PbX_3$ PTFs via fs pulses. **a** First report of red-shifted ASE with **b** a clear pumping threshold feature in spincoated $CH_3NH_3PbI_3$ PTFs [6]. **c** Tunable ASE along visible range in spincoated PTFs via halide substitution [6]. **d** First proof-of-concept lasing in $CH_3NH_3PbCl_xI_{3-x}$ PTF based VCSEL [125]. **e** The first CW-pumped distributed feedback (DFB) lasing operating at 160 K [18]. The cross-sectional scanning electron micrograph (SEM) shows $CH_3NH_3PbI_3$ layer being deposited onto an etched alumina (Al_2O_3) grating. Stable single-mode lasing lasting up to ~1 h under CW pumping was recorded

$\sim$12 μJ cm^{-2} and widely tunable ASE across the various halide substituted films were reported in room temperature conditions of 300 K [6]. Furthermore, lasing was also observed from $CH_3NH_3PbI_3$ single crystals. Deschler et al. demonstrated a prototype VCSEL based on $CH_3NH_3PbCl_xI_{3-x}$ PTFs [125], as shown in Fig. 2.11d. Owing to its versatility as a family of low threshold and emission tunable gain media, it sparked widespread research interest that subsequently aimed at further lowering pumping thresholds and maximising the net gain coefficients. Later in 2017, Jia et al. demonstrated the first CW-pumped DFB lasing, operated at 160 K [18], as shown in Fig. 2.11e. From the scanning electron micrograph presented, the device is constructed with a $CH_3NH_3PbI_3$ film layer deposited onto a thin corrugated alumina Al_2O_3 layer [18]. From the lasing spectra, a corresponding lasing linewidth ~0.25 nm (Q ~ 3120) was obtained and that a far-field spatial profile (not shown here) confirmed its 2nd order lasing emission [18]. By operating near the phase-transiting (162.2 K) temperature of 160 K, local heating imparted by the excitation pulse triggers the back-formation of tetragonal phases kinetically "locked" within the surrounding orthorhombic bulk environment, which in turn creates a carrier-funnelling effect from the orthorhombic phase to the "locked" tetragonal phase. As such, Jia et al. proposed that the lasing originates from the carrier build-up in the "locked" tetragonal phase [18].

Room temperature ASEs in formamidinium lead bromide ($FAPbBr_3$) [126] and $CsPbX_3$ [127] PTFs were also reported in 2016 and 2018, respectively. Specifically, due to the poor solubility of CsX salts in its aprotic solvents, $CsPbX_3$ PTFs are fabricated via a dual-sourced thermal evaporation method [127] instead of convention spin-coating means to prevent incomplete film coverage and pinhole problems [128]. In doing so, an ultralow threshold of $\sim$3.3 μJ cm^{-2} (Fig. 2.12a), an unsaturated net gain coefficient of $g \sim 324\,cm^{-1}$ (Fig. 2.12b, ns pumped) with stable ASE output lasting over 7 h were reported in the $CsPbBr_3$ PTFs [127]. Spectral tunability of ASE in evaporated $CsPbBr_xI_{3-x}$ PTFs are shown in Fig. 2.12c.

In 2019, Pourdavoud et al. introduced thermal imprint (150 °C, 100 bar) onto as-spincoated $CsPbBr_3$ PTFs to form homogeneous surfaces with $\sim$μm grain sizes and complete film coverage [129]. Here, ultralow ASE threshold of $\sim$12.5 μJ cm^{-2} was reported, while DFB and VCSEL lasing thresholds and linewidths of $\left(\sim 7.8\,\mu J\,cm^{-2}, 0.14\,nm\right)$ and $\left(\sim 2.2\,\mu J\,cm^{-2}, 0.07\,nm\right)$ were reported respectively [129]. Interestingly, it was suggested that apart from film morphological optimisation, the co-existence of cubic and orthorhombic $CsPbBr_3$ phases at room temperature led to a natural energy cascade for favourable exciton-transfer processes that aids in ease of accumulation of population inversion [127]. On the other hand, optically smooth and homogeneous $FAPbBr_3$ PTFs were reported by Neha et al. in 2016, where precursors of FABr and $PbBr_2$ are readily dissolved in a mixture of aprotic DMSO + DMF (1:1 vol. ratio) solvents. Here, the presence of mixed DMSO + DMF solvents collectively enhances precursor dissolution and subsequently help to slow down the convective self-assembly during spincoating by slowing down the growth of Perovskite crystallites over longer period of time to foster uniform film morphology [126]. As early as 2015, Stranks et al. demonstrated

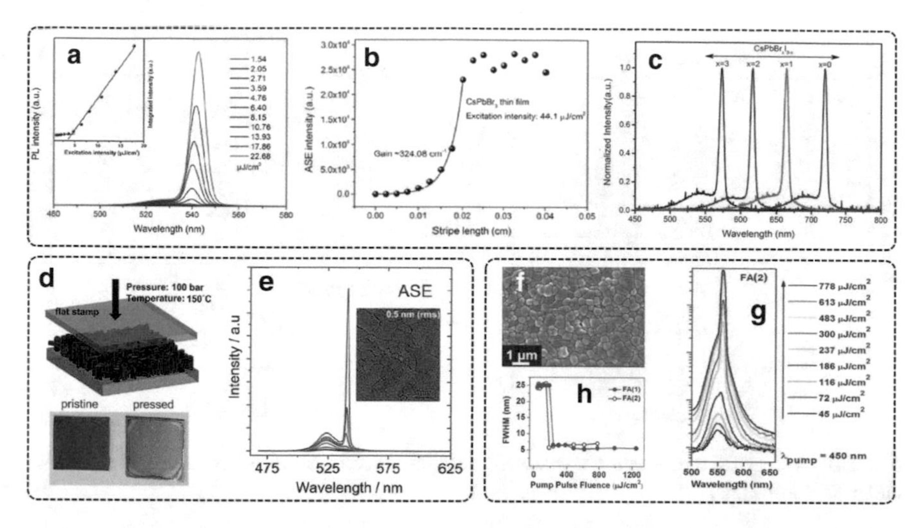

Fig. 2.12 Room temperature ASE observed in $CsPbBr_3$ [127] and $FAPbBr_3$ [126] PTFs. In dual-source thermal evaporated $CsPbBr_3$ PTFs, **a** a red-shifted ASE with threshold of $\sim 3.3\,\mu J\ cm^{-2}$, **b** ns-pumped unsaturated gain coefficient of $g \sim 324\,cm^{-1}$ are reported. **c** ASE tuning via halide substituted groups in $CsPbBr_xI_{3x}$ PTFs. **d** Schematic of thermal imprint stamping of as-spincoated $CsPbBr_3$ PTFs, leading to films with greater surface coverage and optical smoothness. **e** Fluence dependent red-shifted ASE spectra, with inset showing the atomic force microscopic (AFM) image of the thermally stamped film morphology. For $FAPbBr_3$ PTFs, precursors are dissolved in equal volumes of DMSO + DMF. **f** SEM image of a $FAPbBr_3$ PTF spincoated onto mesoporous TiO_2 substrates. **g** Its corresponding fluence dependent ASE spectrum, showing threshold of $\sim 190\,\mu J\ cm^{-2}$ and **h** FWHM narrowing

the potential of PTFs to be applied in areas of wearable optical devices. Essentially, the report showed that polymethylmethacrylate (PMMA) coated $CH_3NH_3PbI_3$ film sandwiched between a gold (Au) metal and cholesteric liquid crystal (CLC) reflector produced the so-called "enhanced ASE" (Fig. 2.13a) as a result of further-narrowing of the ASE linewidth due to etalon waveguiding mechanisms [130]. Under ns-pumping, a reduced threshold of $\sim 7.6\,\mu J\ cm^{-2}$ with enhanced emission stability

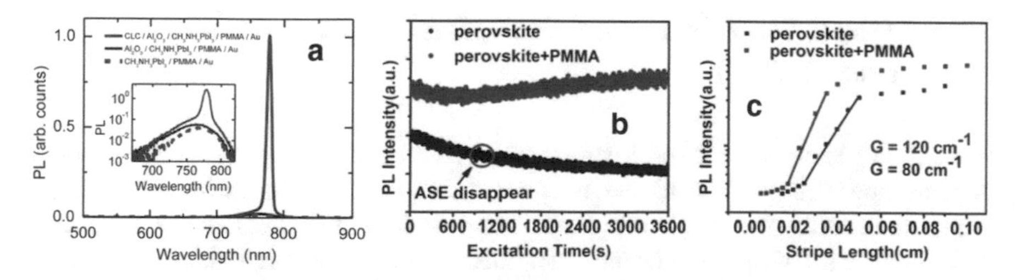

Fig. 2.13 **a** ns pumped "Enhanced ASE" in $CH_3NH_3PbI_3$ PTFs for flexible optoelectronics achieved via optimisation of waveguide in using CLC reflectors and PMMA surface coating [130]. Owing to the passivating effects of PMMA, the "enhanced ASE" becomes **b** more stable over time and possess **c** greater unsaturated gain coefficient than its unpassivated counterpart [131]

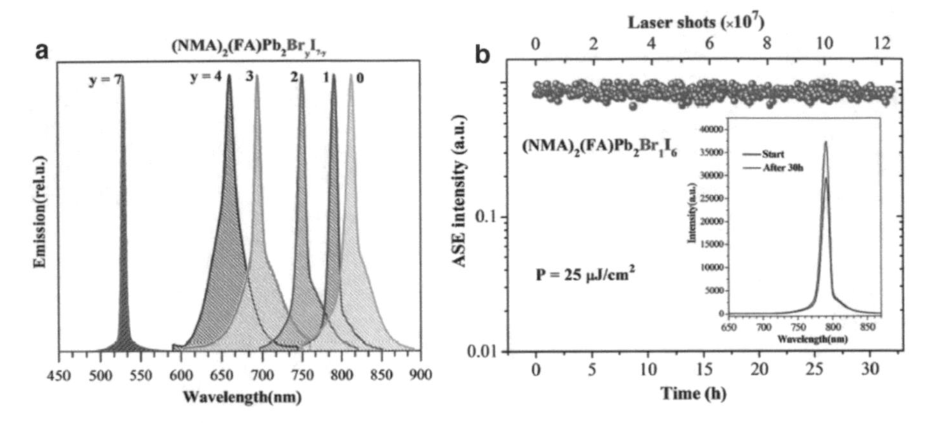

Fig. 2.14 **a** Spectral tuning and **b** stable ASE over prolonged ~fs pulsed excitation in mixed FA-NMA RPP films [30]

(Fig. 2.13b) and unsaturated gain coefficients (Fig. 2.13c) were all observed, as a result of PMMA protection [130, 131]. Greater waveguiding effects leading to modal confinement was also reported separately in "substrate-Perovskite-PMMA" structures than in bare surface unpassivated PTF [131, 132].

Lastly, ASE was also observed in mixed Formamidinium-Napthylmethylammonium (FA-NMA) based RPP films, with ASE pumping thresholds $<20\,\mu J\ cm^{-2}$ and easily tuned across the green-NIR range via halide substitution (Fig. 2.14a), with stable ASE operating up to 30 h (Fig. 2.14b [30]. While the report showcases the robustness of RPP films as stable and low threshold gain media, its underlying light amplification mechanisms and carrier dynamics remains to be further explored. Generally, RPP films are stable materials that have also found other niche applications in solar cells [31, 133], photodetectors [134, 135], FETs [136], LEDs [137–141], lasing [30, 142, 143] and even sensors [144] recently.

2.2.3 Perovskite Micro- and Nanoplatelets (PMPLs and PNPLs)

Perovskite platelets are two-dimensional physical morphologies, which adopts a planar shape with spatial confinement along the crystallite's thickness. Interestingly, this spatial confinement of photoexcited charge carriers and optical modes are responsible for the ease of ASE and lasing observations. Perovskite microplatelets (PMPLs) are micron-sized planar crystallites with sufficiently large dimensions to localise and recirculate optical modes that leads to WGM lasing mechanisms. However, similar localisations cannot be replicated in Perovskite nanoplatelets (PNPLs) as its ~10^1 nm dimensions are unable to support optical feedback for modes resonant along 10^2 nm. Generally, PMPLs and PNPLs can be synthesized in several methods [145]: (i) solution-processing [146–149], (ii) Chemical Vapor Deposition (CVD) [150–152], (iii) solution-vapor phase methods [153] and very recently, (iv) nanoimprinted methods [154]. In the following, readers may find that Perovskite microplates are sometimes termed as "single-crystalline", often due to its similar synthetic approach to its bulk PSC counterparts. To illustrate this, Fig. 2.15 shows an overview of multiphoton pumped ASE and single mode WGM lasing in solution processed PMPLs. Figure 2.15a–d shows the 2PP ASE in inverse temperature crystallised (ITC) $CH_3NH_3PbBr_3$ PMPLs. Unlike the bulk PSC formation, a magnetic stir bar is additionally inserted to disrupt the giant seed crystal formation so as to generate more nucleation sites to favour PMPLs formation with smooth surface and facets [148]. Figure 2.15c shows the onset of 2PP ASE excited by 800 nm fs-pulses beyond the threshold of ~2.2 mJ cm^{-2} [148]. Again, the visually blue-shifted ASE narrowband relative to the PL band is a result of an already relatively red-shifted bulk emission (relative to surface emission) due to reabsorption effects. Yang et al. explains that the ease of 2PP ASE relative to 1PP ASE is due to the former's volume excitation nature instead of the latter's reliance on slow surface photon diffusion during the avalanche stimulated emission stages [148]. Spatially, neighbouring seed

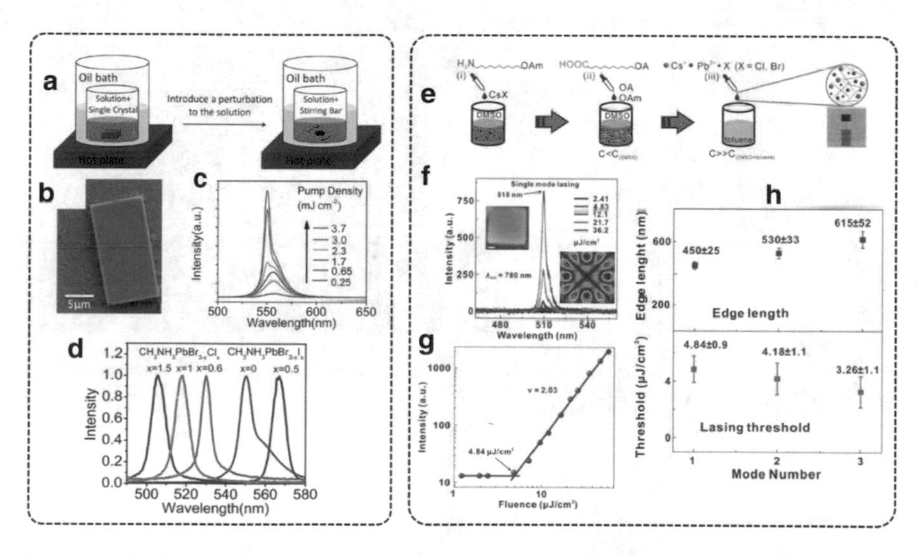

Fig. 2.15 2PP ASE/lasing in solution-processed Perovskites Platelets. **a** Schematic of the stirring-assisted inverse temperature crystallised (ITC) $CH_3NH_3PbBr_3$ PMPLs. **b** SEM image showing smooth surfaces and facets [148]. **c** Pump dependent ASE spectra of $CH_3NH_3PbBr_3$ PMPLs [148]. **d** Tunable ASE via halide substitution in the ITC growth phase. **e** Schematic of the ligand-assisted reprecipitated (LARP) $CsPbX_3$ PMPLs with large edge lengths ~400–800 nm. **f** Pump dependent WGM lasing spectra in ~500 nm sized $CsPbClBr_2$ PMPLs at 77 K, where a simulation of its modal electric field distribution confirms WGM lasing mechanisms. **g** verification of 2PP induced lasing via lasing intensity scaling quadratically with pumping fluence. **h** The statistical relationship between number of lasing modes, microplatelet size and thresholds [146]

photons within its volume excitation can take part to seed the avalanche stimulated emission [148]. Figure 2.15d illustrates the robustness of tunable upconverted ASE in these "single-crystalline" PMPLs. Single mode WGM lasing in Ligand-Assisted Re-Precipitated (LARP) $CsPbClBr_2$ PMPLs has also been observed by Huang et al., as shown in Fig. 2.15e–g [146]. Contrary to the ITC approach, the LARP approach relies on stringent surface chemistry via ligands, where the swift injection of dissolved precursors into toluene in the presence of Oleylamine (OAm) and Oleic acid (OA) ligands allow for the formation of large square $CsPbX_3$ PNPLs with edge lengths of ~400–800 nm. Figure 2.15f shows the onset of WGM lasing in $CsPbClBr_2$ under 2PP regime at 77 K conditions, with a threshold of $\sim 5\,\mu J\,cm^{-2}$ (and $\sim 60\,\mu J\,cm^{-2}$ at 293 K) [146]. Its corresponding resonant modal electric field distribution is shown in the right inset, which further affirms WGM lasing mechanisms [146]. Figure 2.16g confirms the evidence of 2PP excitation induced lasing due

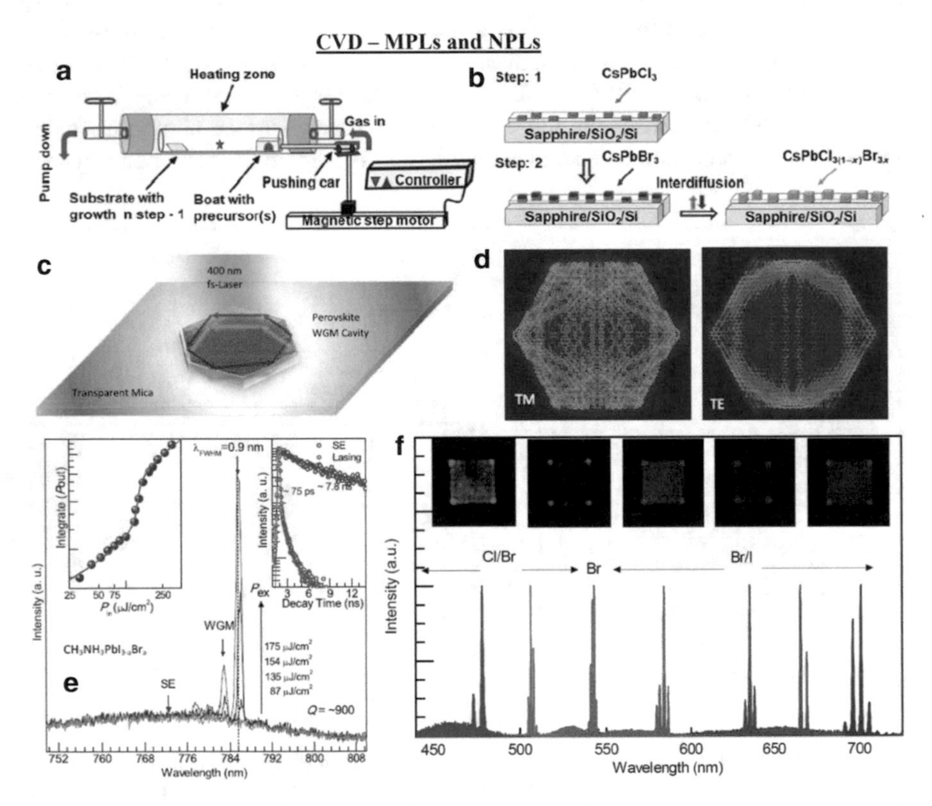

Fig. 2.16 Schematic of **a** ternary and **b** mixed halide quaternary PMPLs grown via one and two-step CVD, respectively [150]. **c** Schematic illustrating the optical feedback path of WGM lasing mechanisms in hexagonal $CH_3NH_3PbI_3$ PMPLs and **d** its respective simulated TE and TM waveguided modal electric field distributions. **e** Pump dependent WGM lasing spectra, with inset showing threshold feature (left inset) and drastic carrier lifetime shortening once above lasing threshold excitation. **f** Spectrally tunable WGM lasing in CVD grown PMPLs on mica substrates, characteristically waveguided along TM modes [152]

to the quadratic slope relation between lasing intensity with respect to the pumping fluence, where $\nu = \frac{\log I_{em}}{\log I_{exc}}$. Lastly, Fig. 2.15h illustrates the statistical relation between a PNPL's edge length, number of lasing modes and its corresponding threshold value. As expected, the larger the PNPL's size, the greater the number of participating lasing modes with lower corresponding lasing thresholds. Huang et al. explains that while this shows that multimodal WGM lasing is easier achieved in larger sized PNPLs due to lower thresholds, but the main interest is still to pursue single mode operation in nanoscale structures to match the purpose of miniaturization and stabilization of devices. Finally, highly polarised transverse electric (TE) and magnetic (TM) WGM lasing in RPP microplatelets have also been observed and was first reported by Li et al., possessing ultralow threshold of $\sim 8\ \mu J\ cm^{-2}$ with robust lasing stability up to ~10 h, under conditions of room temperature ~fs pulsed excitation conditions [147].

Chemical Vapor Deposition (CVD) is another often employed technique to reliably synthesize high quality two-dimensional materials with controllable sizes and thicknesses [145]. Thus, CVD grown PMPLs have also yielded positive results and hold the advantage to be easily transferrable into on-chip integrated photonic circuits for lasing application, where exploration into potential electrically driven Perovskite platelet WGM laser devices can open doors to applications such as optical sensing [155]. In Zhang et al.'s report on WGM lasing in PMPLs, the samples are in fact, CVD grown [151]. A schematic describing the CVD growth of ternary PMPLs and its mixed halide quaternary variants are shown in Fig. 2.16a and b, respectively [150]. Here, Hossain et al. shows that the mixed halide $CsPbX_{3-m}Y_m$ can be acquired simply through sequential two-step alloy growth of its ternary constituents followed by interdiffusion within the CVD chamber tube [150]. Interestingly, early reports of $CH_3NH_3PbI_{3-x}Y_x$ PMPLs grown on muscovite mica substrates also showed that tuning the halide component would also tune its resultant shape, from triangular to hexagonal [151]. Figure 2.16c and d illustrates the optical feedback path leading to WGM lasing in $CH_3NH_3PbI_3$ PMPLs as well as its corresponding TE and TM modal electric field distribution profiles, respectively [151]. Under ~ fs pulsed excitations, lasing thresholds ranging from $\sim 37-128\ \mu J\ cm^{-2}$ were observed in these hexagonal $CH_3NH_3PbI_3$ PMPLs, as shown in Fig. 2.16e. Similar efforts to grow uniformly square-shaped $CsPbX_3$ PMPLs tunable across the visible range with average Q-factor reaching as high as ~4800 has been reported, as shown in Fig. 2.16f. In this work, $CsPbX_3$ PMPLs with edge length spanning from ~9 to 29 μm and lasing threshold down to $\sim 2\ \mu J\ cm^{-2}$ has been observed [152]. Atomically smooth triangular $CH_3NH_3PbI_3$ PMPLs with lower WGM lasing thresholds and higher Q-factor have been proposed, where initial CVD-grown triangular PbI_2 templates on mica substrates are subsequently converted into triangular shaped $CH_3NH_3PbI_3$ PMPLs with surface roughness lesser than 2 nm [156]. Interestingly, another form, called polariton lasing has been observed in CVD-grown PNPLs by Rui et al. in 2017 [157]. Here, it was reported that these $CsPbCl_3$ PNPLs could produce ultralow polaritonic lasing thresholds of $\sim 12\ \mu J\ cm^{-2}$ that emits in the deep-blue range [157]. As we will see in Chap. 3, the polaritons are coupled light-carrier quasi-particles that produces coherent emission via polaritonic condensate instead of the usual photonic population inversion mechanisms which in turn accounts for much lower lasing thresholds than photon lasing.

Another report has also shown a similar "two-step" solution-vapor phased method for synthesizing hexagonal $CH_3NH_3PbI_3$ microdisks (PMPLs), as shown in Fig. 2.17a [153]. In the first step, a preliminary $CH_3NH_3PbI_3$ PTF is acquired via conventional spin-coating means and followed by the second step, in which the $CH_3NH_3PbI_3$ PTF is transferred into a horizontal tube reaction chamber to facilitate its MPLs self-assembly mechanisms [153]. Figure 2.17b shows that under Q-switched ~ns pulsed excitation, low ASE pumping thresholds along several tens of ~μJ cm^{-2} can be achieved. Figure 2.17c shows that multimodal WGM lasing with similarly low thresholds are obtained under confocal microscopic excitations. Both measurements showcase that solution-vapor phased synthesis is possibly another reliable synthetic route for engineering robust PMPL-based lasers.

Preliminary attempts to construct $CH_3NH_3PbX_3$ PMPL-based laser arrays was achieved via NIL for both multimodal [158] and single-modal WGM lasing operations [154, 158]. Figure 2.18a and b shows a schematic (left) and a close-up false coloured SEM image of the actual fabrication, as well as limited spectral tunability along green-NIR ranges, respectively [154]. Ultralow lasing thresholds in both ~fs pulsed and ~ns pulsed excitations at 7 μJ cm^{-2} and 150 μJ cm^{-2} were observed, respectively [154]. Such results highlights the potential path forward in developing reproducible and uniformly sized and edge-to-edge smooth microdisk arrays suitable for integration into on-chip devices [154]. In particular, the ability to control the disks' edge roughness is important because from a device standpoint, edge roughness

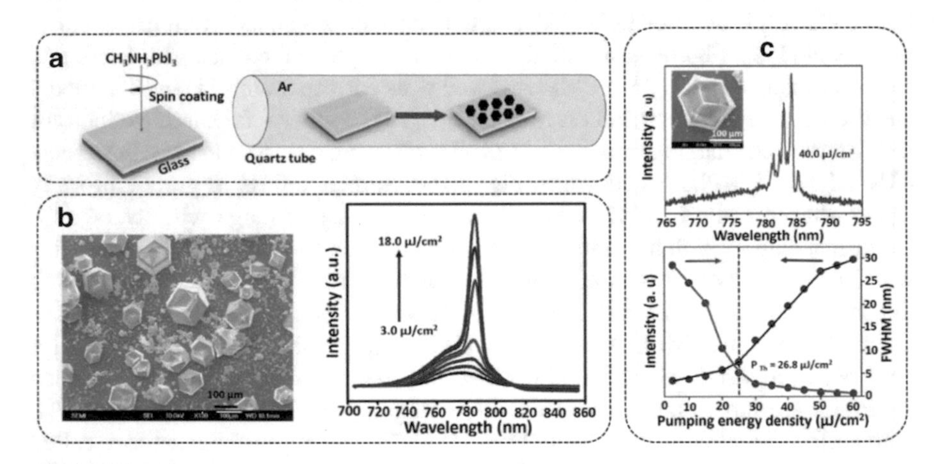

Fig. 2.17 Evidence of ASE and lasing in self-assembled $CH_3NH_3PbI_3$ microdisks [153] via **a** the solution-vapor synthetic route. **b** Its SEM image and fluence dependent ASE characterisation under ~ns pulsed excitation. **c** WGM lasing excited via micro-PL, showing an equivalently low lasing threshold to its ASE threshold

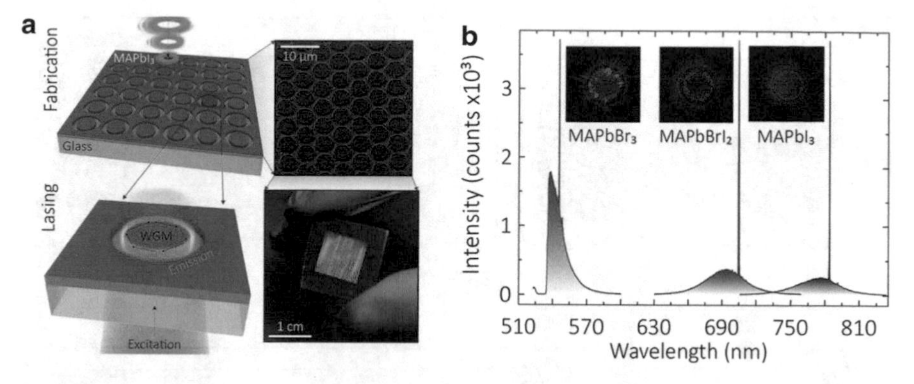

Fig. 2.18 Single-mode WGM lasing operation of $CH_3NH_3PbX_3$ microdisks synthesized via **a** nanoimprint lithography assisted with laser ablation (NIL) [154]. **b** Single-mode lasing in $CH_3NH_3Pb(Br/I)_3$ microdisks under sub-ns pulsed excitations with thresholds along 10^2 µJ cm^{-2}. **c** Size-dependent single-mode lasing in the $CH_3NH_3PbI_3$ microdisks

significantly impacts the WGM recirculation and inhibit side modes from participating in laser oscillations. Here, an unsaturated gain coefficient ~200 cm^{-1} is estimated, which is consistent with other experimentally determined single-crystalline morphologies. In a similar attempt, Feng et al. also successfully created a WGM based microlaser array consisting of tessellated microplates with uniform size and precise structural localisation using an interesting "liquid-knife" strategy [159].

2.2.4 Perovskite Micro- and Nanowires (PMWs and PNWs)

Perovskite wires are one-dimensional physical morphologies, which adopts a rod-like shape with two degrees of spatial confinement. Fabry–Perot lasing can easily occur in Perovskite microwires (PMWs) and nanowires (PNWs) due to longitudinal optical feedback occurring between the rod structure's end facets, repetitively (see Chap. 1 Fig. 1.4). Interestingly, WGM lasing can also occur in PMWs, where transverse optical feedback occurs along the rod's circular cross-section instead [160]. Similar to its platelet counterparts, PMWs and PNWs possess strong anisotropy [145], which makes them effective materials that can be applied in biosensing, ultra-resolution imaging, catalysis and medical diagnosis [161, 162]. Generally, PMWs and PNWs are also solution-processed [55, 59, 163] or CVD (vapor-phase) [164] grown, either directly or post-synthetically transformed [145].

In 2015, Zhu et al. was the first to demonstrate record-low Fabry–Perot lasing with threshold down to ~600 nJ cm^{-2} in solution-processed $CH_3NH_3PbX_3$ PNWs [55], as shown in Fig. 2.19a, b. In the insets, SEM images revealed smooth surfaces and facets, which serve as low-loss high Q-factor that contributes greatly to the observation of ultralow lasing thresholds [55]. Furthermore, bright resonances localised at both end-facets shown in its optical images confirm longitudinal waveguiding effects

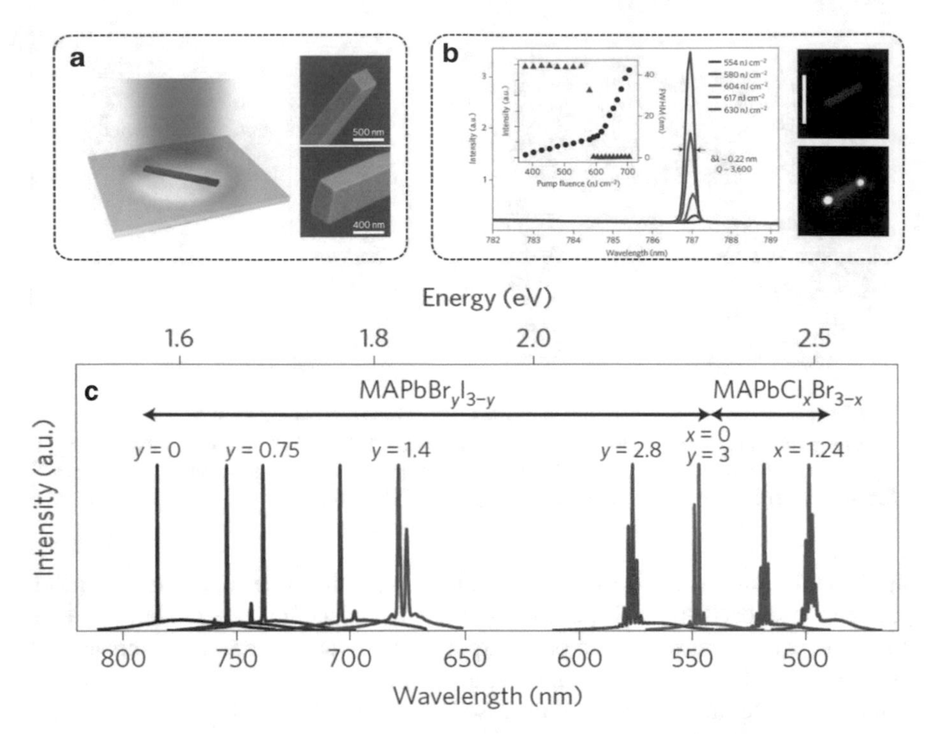

Fig. 2.19 Fabry–Perot lasing in $CH_3NH_3PbX_3$ PNWs. **a** A schematic of the home-built far-field micro-PL excitation geometry, with inset showing SEM image of the PNWs. **b** The pump dependent lasing spectra of $CH_3NH_3PbI_3$ PNWs, showing record low 600 nJ cm^{-2} lasing threshold with Q $\sim$ 3600. The inset show optical images recorded when the PNW was pumped below (top) and above (bottom) lasing threshold. **c** $CH_3NH_3PbX_3$ PNW Laser's emission tunability upon halide substitution

arising from Fabry–Perot lasing mechanisms [55]. Figure 2.19c shows the halide-substituted partial lasing emission tunability in $CH_3NH_3PbX_3$ PNWs, ranging from turquoise to NIR [55]. Similar lasing actions were also observed in CVD grown $CH_3NH_3PbI_3$ PNWs of $\sim$20 μm long [164] and low threshold ($\sim$674 μJ cm^{-2}) 2PP lasing in solution-grown $CH_3NH_3PbBr_3$ PMWs [59]. Selective Fabry–Perot lasing while suppressing inevitable competition from WGMs have been achieved by employing higher refractive index substrates to prevent total internal reflection induced waveguides [162]. In this work, Wang et al. reports uniform microlaser arrays based on $CH_3NH_3PbBr_3$ PMWs embedded on silicon chips [162].

For the Formamidinium variants, it was determined that CH_3NH_3Br can be used as a stabilising additive to improve the crystallinity of $FAPbI_3$ PNWs after thermal conversion, and would lead to lower lasing thresholds [163]. Figure 2.20a shows the onset of lasing in a CH_3NH_3Br-$FAPbBr_3$ PNW, where the bottom and top inset shows the respective optical images when the PNW is excited below and above threshold. Again, the end-facet resonances verify Fabry-Perot lasing mechanisms. Interestingly, the work also compared and demonstrated the inverse relationship

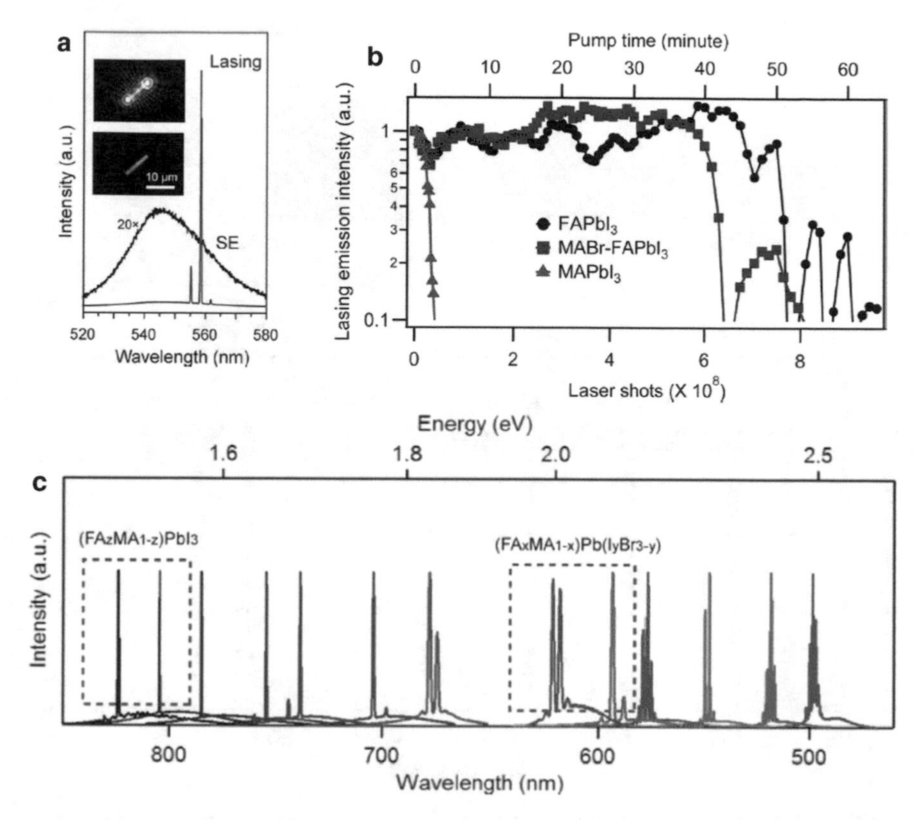

Fig. 2.20 Fabry–Perot lasing in CH_3NH_3Br-$FAPbX_3$ PNWs [163]. **a** Emission spectra of $FAPbBr_3$ NW below and above lasing threshold excitations. Inset shows optical images of the excited PNW at above (top) and below (bottom) thresholds. **b** A comparison of lasing photostability in PNWs. **c** Tunable lasing emission in mixed cationic (MA/FA)$PbX_{3-m}Y_m$ NWs

between lasing thresholds recorded against its lasing stability in both $CH_3NH_3PbI_3$ and $FAPbI_3$ PNWs. As shown in Fig. 2.20b, while pure $FAPbI_3$ NWs suffered from higher lasing threshold, it possessed longer photostability than $CH_3NH_3PbI_3$ PNWs, and the converse is also true [163]. Lastly, Fig. 2.20c shows the other degree of lasing emission tunability in PNWs by combining both cationic and halide substitutions [163]. In this work, Fu et al. proposed to explore the feasibility of double cationic mixed (MA/FA)$PbX_{3-m}Y_m$ that can also aid in lowering photoexcitation induced ion-migration [165] effects [163].

Widely tunable Fabry–Perot lasing via vapor-phased converted $CsPbX_3$ PNWs are also subsequently observed, as shown in Fig. 2.21. Essentially, a direct synthesis of mixed halide systems in such fully inorganic systems are challenging, due to it being thermodynamically favourable to form a δ-phased $CsPbI_3$ side-product. A schematic of the vapor-conversion synthesis is shown in Fig. 2.21a. Interestingly, they showed Fabry–Perot lasing in deep-blue emitting $CsPbCl_3$ NWs, which has yet to be seen

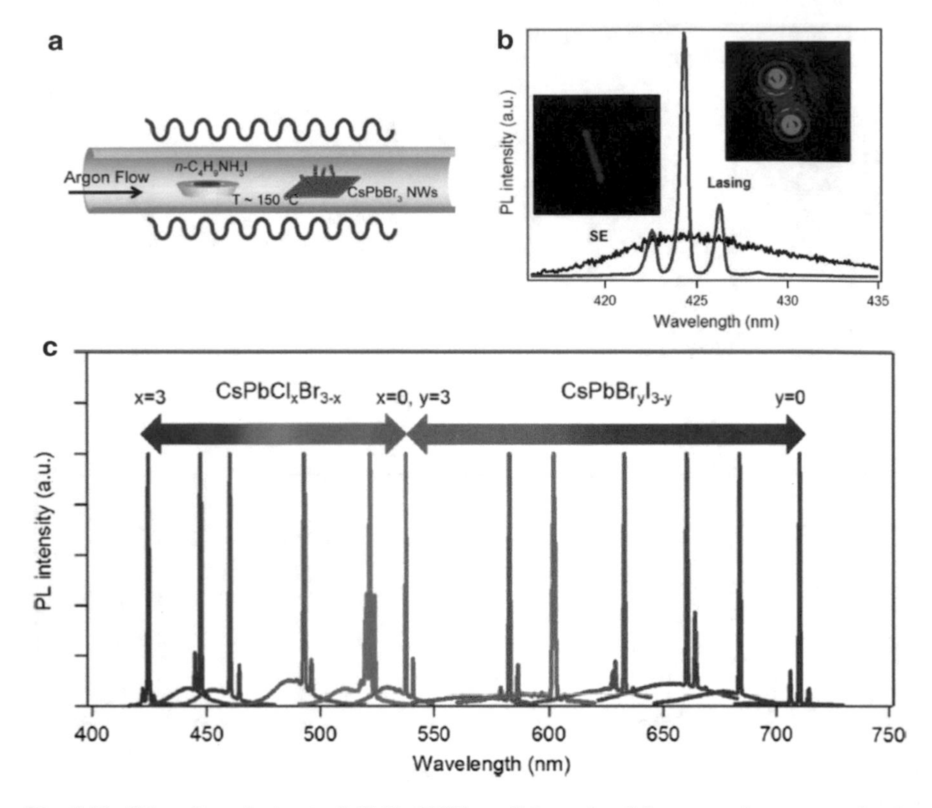

Fig. 2.21 Fabry–Perot lasing in $CsPbX_3$ PNWs. **a** Schematic of the vapor-phase conversion in synthesizing $CsPb(Br/I)_3$ NWs. **b** Emission spectra of $CsPbCl_3$ deep blue-emitting below and above lasing threshold excitations. Inset shows the end-facet resonances characteristic of axial waveguiding that gives rise to Fabry–Perot laser mode oscillations. **c** The $CsPbX_3$ NW laser wavelength tuning based on halide substitution

in the hybrid (MA and FA variants) PNWs. Lastly, similar halide-substituted lasing emission tuning is shown in Fig. 2.21c.

Lastly, Perovskites nanowires are excellent testbeds for studying polaritonics, due to its their naturally high oscillator strengths and strong photon confinement induced by the strong material anisotropy [166]. Thus, intensive efforts have been placed into uncovering the relative interplay between exciton-photon coupling [166–169] and exciton–polariton coupling [168, 170] that effectively leads to photonic lasing instead of polaritonic lasing. Alternatively, plasmons [171] have also been suggested to be the underlying mechanism responsible for the direct observation of Fabry–Perot lasing in these PMWs and PNWs. Due to the scope of this Section, we shall revisit the various debated lasing/optical gain mechanisms in these Perovskite wire-like systems in Chap. 3.

2.2.5 *Perovskite Micro- and Nanocrystals (PMCs and PNCs)*

With reference to Fig. 2.7b, the Perovskite micro- (PMCs) and nanocrystalline (PNCs) morphologies are also commonly employed in optical gain studies, comprising up to nearly a-third of the entire compiled reports. Similarly, PMCs are also often referred as "single-crystalline" microcubes/cuboids due to its similarity towards its bulk single crystalline counterparts in terms of synthetic routes and properties [172]. In 2016, early reports of PMC lasing often demonstrated frequency upconverted WGM lasing under multiphoton excitations [173–175]. For instance, single mode Fabry–Perot lasing arising from blue-emitting $CsPbCl_3$ PMCs excited using 2PP up to 6PP was first observed by Yang et al., albeit at 215 K, as shown in Fig. 2.22 [175]. Notably, each regime's non-linear absorption cross-sections were measured to be extremely large, which contributed to the ease of lasing buildup.

WGM lasing mechanisms are also naturally attainable due to total internal reflection allowed along its cross-sections [6, 153], as illustrated in Fig. 2.23a [176]. Depending on the excitation geometry (not shown here), the choice of macroscopic and microscopic excitations can lead to ensemble ASE or single PMC WGM lasing, respectively [153]. In a report by Zhou et al., a $CsPbBr_3$ PMC with edge length ~2.2 μm demonstrated single mode WGM lasing with a threshold of ~17 μJcm^{-2} and a corresponding Q-factor ~8500 under ~fs pumping condition [176]. In addition, similar lasing was observed even under ~ns laser excitation conditions of 355 nm, 1.1 ns and repetition rate of 20 kHz, despite imparting more thermal effects (local

Table 1. Multiphoton Absorption Coefficients of $CsPbCl_3$ Microcrystals at Room Temperature

Multiphoton Excitation	2PA (cm/GW)	3PA (cm^3/GW^2)	4PA (cm^5/GW^3)	5PA (cm^7/GW^4)	6PA (cm^9/GW^5)
Wavelength (nm)	800	1200	1600	1800	2200
Intensity I_o (GW/cm^2)	4.7	12	57	90	160
α_n	3.8	0.089	1.1×10^{-4}	2.3×10^{-7}	1.1×10^{-10}

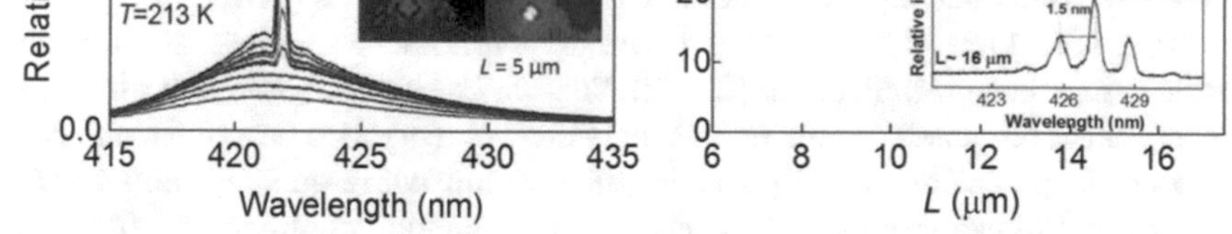

Fig. 2.22 Single-mode Fabry–Perot Lasing at the end-facets of an isolated $CsPbCl_3$ PMC end facets. Top inset shows the SEM image of $CsPbCl_3$ PMCs. The table summarises the huge optical non-linear absorption constants from 2 to 6PA regimes [175]

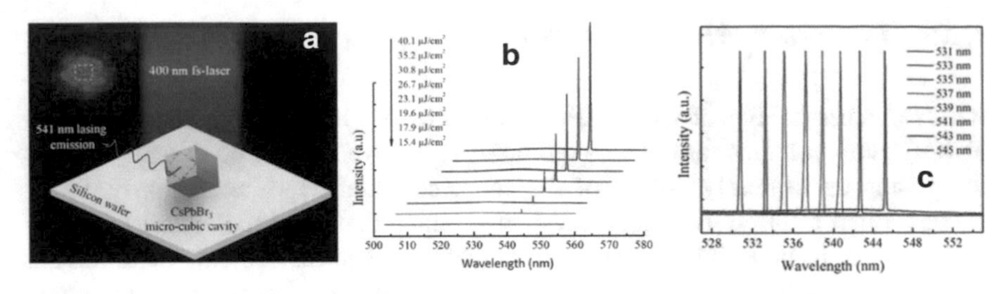

Fig. 2.23 WGM Lasing in $CsPbBr_3$ PMCs [176]. **a** Schematic of WGM lasing under single photon pumping regime. **b** Fluence-dependent lasing spectra of an isolated $CsPbBr_3$ PMC and **c** size-dependent $CsPbBr_3$ PMCs leading to finetuning of green lasing peak

temperature and increased lasing thresholds) to the PMC [176]. Interestingly, size-dependent tunable WGM lasing finetuned along 530–545 nm were found possible over a range of $CsPbBr_3$ PMC edge lengths ~1.1 to ~2.4 μm [176]. Generally, the relation between resonant lasing position and edge length can be expressed as [177]:

$$2\sqrt{2}\frac{nL}{m} = \lambda_m \tag{2.3}$$

where n, L, m are the material refractive index, edge length and mode order number while λ_m is the corresponding mode resonant wavelength for lasing. Especially for smaller sized microcavities that supports single mode, its mode number m remains unchanged and the resonant lasing position λ_m is seen to scale linearly with edge length L [176].

Recently in 2019, VCSEL lasing in RPP based $(PEA)_2Cs_{n-1}Pb_nBr_{3n+1}$ PMC microcavities were also reported, with relatively low ns-pumped lasing threshold at $\sim$500 μJ cm^{-2} [143]. In this work, Zhai et al. attributed the success to the excellent passivation of surface trap states through long chain Phenyl–ethyl ammonium ligands (PEA). One interesting way of further lowering lasing thresholds in PMCs was through the introduction of detachable aluminum nanoparticle based substrates in an effort to promote stronger modal spatial confinement and near-field enhancements along the metal-Perovskite (dielectric) surface to combat optical losses. Here, Chen et al. found a ~27% reduction in lasing threshold accompanied by an ~tenfold increase in stimulated emission rates in $CH_3NH_3PbBr_3$ PMCs [178].

PNCs, with their high PLQY and strongly excitonic properties, are attractive candidates for lasing applications [25, 52]. Strictly, one should avoid using the terms nanocrystals and quantum dots interchangeably. Nanocrystals describe small crystallites with dimensions usually smaller than 10 nm while quantum dots refer to extremely small nanocrystals that receives strong quantum confinement effects as a result of having dimensions comparable or smaller than its characteristic exciton bohr radius. For convenience of reference, several useful constants including the exciton bohr radii for common lead halide Perovskites are listed in Table 2.3. The terms E_g, E_B^X, r_b, ϵ_∞, $\langle m_h^* \rangle$ and $\langle m_e^* \rangle$ refers to the material bandgap, excitonic binding energy,

Table 2.3 Summarised table of constants in both Fully-Inorganic and Hybrid PNCs

	E_g/eV	E_B^X/meV	r_b/nm	ϵ_∞	$\langle m_h^* \rangle$	$\langle m_e^* \rangle$
Fully-inorganic Perovskite						
$CsPbCl_3$	3.02	72	2.5	4.07	0.17	0.20
$CsPbbr_3$	2.20	38	3.5	4.96	0.14	0.15
$CsPbbI_3$	1.64	20	6.0	6.32	0.13	0.11
Hybrid organometallic Perovskite						
$CH_3NH_3PbCl_3$	2.88–3.10	50–69	–	–	0.47	0.34
$CH_3NH_3Pbbr_3$	2.30	40–76	2.0	3.29	0.23	0.21
$CH_3NH_3PbI_3$	1.60	13–50	2.2–2.8	6.5	0.29	0.23

excitonic bohr radius, dielectric constant and effective hole and electron masses, respectively.

One major advantage of PNC emitters is its lasing emission tunability. Apart from direct halide substitution in early synthetic preparations, post-synthetic anionic exchange treatments have been shown possible due to their colloidal nature, with the cited advantage of retaining "parent" ensemble size-distributions and physical shapes with excellent reproducibility [51, 179–182]. In practice, such versatility presents PNCs as an attractive class of Perovskite lasing materials that offers a spectral coverage broader than the National TV System Committee (NTSC) display standards, as shown in Fig. 2.24. Generally, PNCs are synthesized either via (I) Ligand Assisted RePrecipitation (LARP) [24] or (II) Hot-injection methods [24, 183], depending on the chosen precursor salts' degree of solubility in aprotic solvents. $CsPbX_3$ PNCs adopts the hot-injection route (Fig. 2.24a) as CsX salts are much less soluble than PbX_2 salts in aprotic solvents like DMF and DMSO. Thus, the Cs-source and PbX_2 salts must be separately dissolved, with the former provided by Cs_2CO_3 that is initially reacted into a Cs-Oleate. Next, the Cs-Oleate will then be swiftly injected into the reaction flask containing dissolved PbX_2 salts in the presence of ligands such as Oleylamine (OLA) and Oleic Acid (OA). On the other hand, since both CH_3NH_3X and PbX_2 salts are extremely soluble, they can be homogeneously dissolved in its aprotic solvents and subsequently directly injected into the non-polar "bad" solvent such as toluene (LARP, Fig. 2.24d). While LARP seems more convenient based on these descriptions, it has several disadvantages in comparison to PNCs fabricated by hot-injection. Firstly, remnant unreacted aprotic solvents can easily attack the PNCs and degrade its overall colloidal stability and cause the material's lasing thresholds to increase arbitrarily [184, 185]. In fact, a study revealed that common solvents like DMSO and THF form coordinating complexes with the PbI_2 precursor salts that result in PNCs with defective surfaces and volume vacancies, while γ-butyrolactone and acetonitrile are found to be better and noncoordinating alternatives [186]. Secondly, DMF and DMSO are toxic, thus making LARP an unsafe route when considering up-scaling for future productions [186].

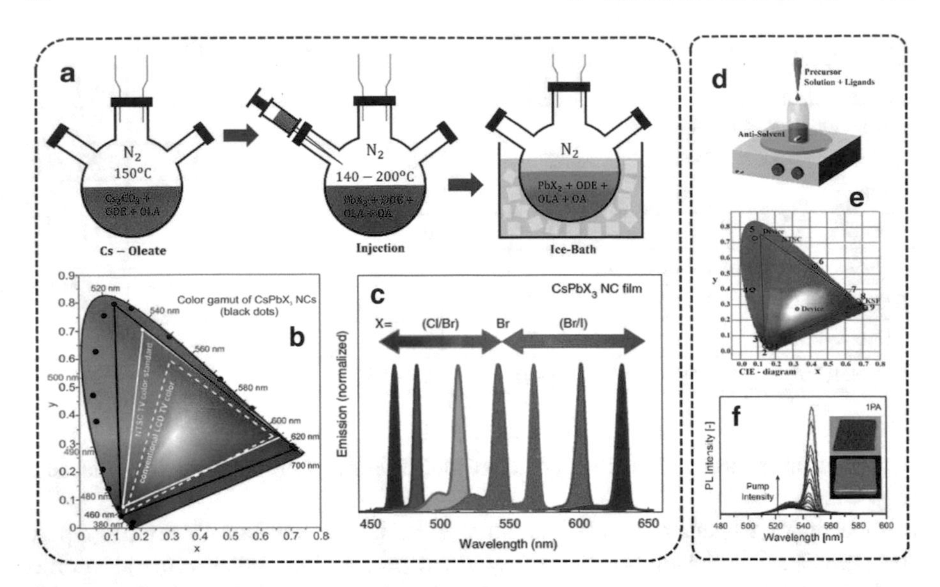

Fig. 2.24 Synthetic protocols, colour gamuts and ASE spectra in fully-inorganic and hybrid Perovskites. **a–c** The hot-injection schematic, tunable emission coverage [183] and ASE spectra of $CsPbX_3$ PNCs [52]. **d–f** The LARP schematic, tunable emission coverage [24] and ASE spectra of $CH_3NH_3PbBr_3$ PNCs [25]. Generally, both classes of PNCs provide spectral coverages even wider than NTSC display standards [38, 53]

Other post-synthetic anionic exchange routes include the so-called immobilised "dry" solid-state reaction of PNCs via potassium halide salt (KX) matrices [180] and also photo-induced reductive dissociation of PNC halide ions using various dihalomethane [182], although tunable ASE and lasing characterisations remain to be verified. Nevertheless, they imply that PNCs hold great potential for tunable laser wavelength division multiplexing (WDM) for optical communications and network [187].

Interestingly, several studies were conducted, mostly to provide alternative pathways for tailored applications. For instance, Imran et al. introduced the use of benzoyl-halides instead of toxic PbX_2 salts and to allow halide-tuning without worrying about lead stoichimetries during hot-injection [188]. Mainly, this approach allows for the formation of near-unity PLQY $CsPbX_3$, $CH_3NH_3PbX_3$ and $FAPbX_3$ PNCs with comparable ASE thresholds down to several $\mu J\ cm^{-2}$, as seen in Fig. 2.25 [188]. Another alternative to conventional PNC fabrication strategy, is the so-called "in-situ" technique, where PNCs are fabricated on the spot with the intention to confine the ensemble size within the growth process directly, allows for large-scale production and has shown excellent photo- and electroluminescence properties [189], all of which bodes well for electrically pumped lasing studies. This approach ranges from molecular ligand engineering [116, 190–192] and in-situ fabrication of PNCs embedded in polymer [193–195] and crystal matrices [196, 197]. Owing to its wide range of synthetic approach, the "in-situ" fabricated PNC films can find applications

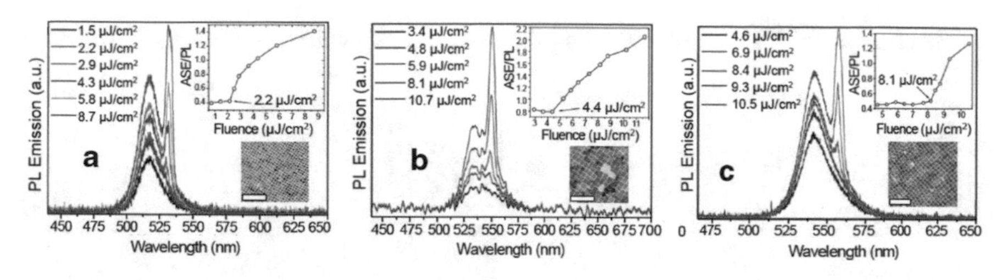

Fig. 2.25 ASE of **a** $CsPbBr_3$ PNCs, **b** $CH_3NH_3PbBr_3$ PNCs and **c** $FAPbBr_3$ PNCs synthesized via hot-injection using Benzoyl halide precursors [188]. Inset shows the respective synthesized ensembles under TEM

in back-light units for liquid crystal displays (LCD), integration into silicon detectors for enhanced UV-detection and most importantly electrically pumped PNC lasing by virtue of its strong electroluminescence [189].

PNCs also found their foothold in upconversion ASE and lasing, similar to its PMCs counterpart and is due to its giant optical non-linearity. Generally, frequency upconverted lasers serve in areas of non-invasive/destructive biomedical photonics [57]. For instance, open aperture Z-scan measurements revealed a giant two-photon absorption cross-section of $\sigma_{abs}^{2PA} \sim (1.0 \pm 0.2) \times 10^5$ GM in BnOH-treated $CH_3NH_3PbBr_3$ colloids [25] and $\sigma_{abs}^{2PA} \sim 1.2 \times 10^5$ GM in $CsPbBr_3$ colloids [198], as shown in Fig. 2.26. These σ_{abs}^{2PA} values are in fact larger than state-of-the-art red-emitting CdSe QDs [199, 200] by an order of magnitude and contributes to ease of acquiring population inversion, thereby relatively reducing 2PP ASE thresholds. For reference, 2PP ASE thresholds of $\sim$570 μJ cm^{-2} and $\sim$2.5 mJ cm^{-2} in BnOH-treated $CH_3NH_3PbBr_3$ [25] and $CsPbBr_3$ [198] PNC films. Very recently, a large $\beta \sim 0.76\,\text{cm/GW}$ accompanying an ultralow 2PP ASE and microcapillary WGM lasing thresholds of $\sim$300 μJ cm^{-2} and $\sim$310 μJ cm^{-2} was also reported in $FAPbBr_3$ PNCs too [201].

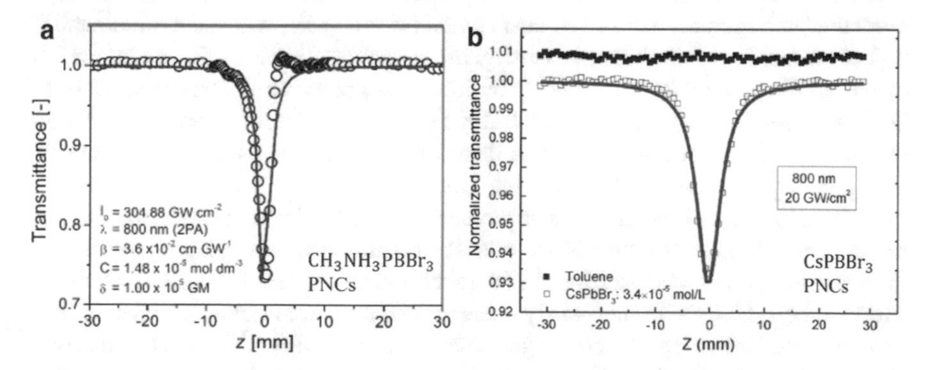

Fig. 2.26 Giant two-photon absorption cross-sections in **a** $CH_3NH_3PbBr_3$ and **b** $CsPbBr_3$ colloidal PNCs measured using the open-aperture Z-scan

Generally, σ_{abs}^{2PA} is calculated using the position dependent transmission function [202]:

$$T(z) = \sum_{m=0}^{\infty} \frac{\left[\frac{-q_0}{1+\left(\frac{z-a}{z_0}\right)}\right]^m}{(m+1)^{3/2}} \tag{2.4}$$

where z and z_0 are the scanning position relative to beam focus ($z = 0$) and the Rayleigh range, respectively. As seen in Fig. 2.26, fitting the open-aperture Z-scan data with Eq. (2.4), usually with up to the third order ($m = 3$) yields a value of q_0, which allows us to next calculate the two-photon absorption coefficient $\beta = \frac{q_0}{I_0 L}$, where L is the cuvette thickness and $I_0 = \frac{2P_0}{\sqrt{\pi} f_{rep} \pi w_0^2 \tau_{exc}}$ is the excitation pulse's peak intensity [202]. Lastly, σ_{abs}^{2PA} can be determined as follows [202]:

$$\sigma_{abs}^{2PA} = \frac{hc\beta}{\lambda C n_A} \tag{2.5}$$

where λ, C and n_A are the excitation wavelength, colloidal concentration, and Avogadro's constant, respectively.

2.3 Summary and Conclusion

In this chapter, we introduced the multitude of structural and morphological dimensionalities that the Perovskite material can take. Structurally, Perovskites follow a chemical formula of $A_{4-m}BX_{6-m}$, where m is its structural dimensionality and dictates the number of directions in which the BX_6^{4-} octahedra could tessellate. Next, we considered the various optical properties of Perovskites that make them attractive gain media, such as low cost and facile synthesis, wide spectral tunability and coverage, low gain thresholds and giant optical nonlinearity. Following on, we discussed the general trajectory of Perovskite lasing research starting from 2014 till 2020, recognising that the end goal in this field is to construct a room-temperature electrically driven Perovskite laser; realising and optimising CW pumped Perovskite lasing while proposing various heat-sink strategies. In Sect. 2.2, we presented a broad overview of ASE and lasing that is observed in all kinds of Perovskite morphologies, ranging from bulk Perovskite single crystals to nanocrystallites. Lasing reports are limited in bulk PSCs due to the inherent competition between bulk volume and surface carrier effects. PTFs have seen vastly reduced ASE thresholds and boosted gain coefficients when correcting film coverage and rectifying grain-boundaries. Lasing could also be achieved with optical feedback occurring between end facets in microstructures without requiring any external reflective surfaces. Perovskite nanostructures have demonstrated ultralow ASE thresholds at sub μJ cm^{-2} orders, accompanied with strong optical nonlinearity that sets them high on the list of gain media for frequency up-converted lasing applications.

References

1. A. Kojima, K. Teshima, Y. Shirai, T. Miyasaka, Organometal halide perovskites as visible-light sensitizers for photovoltaic cells. J. Am. Chem. Soc. **131**(17), 6050–6051 (2009). https://doi.org/10.1021/ja809598r
2. W.S. Yang et al., Iodide management in formamidinium-lead-halide-based perovskite layers for efficient solar cells. Science **356**(6345), 1376–1379 (2017)
3. N. NREL, *Best Research-Cell Efficiencies* (National Renewable Energy Laboratory, Golden, Colorado, 2019)
4. N.G. Park, Research direction toward scalable, stable, and high efficiency perovskite solar cells. Adv. Energy Mater. **10**(13), 1903106 (2020)
5. M.A. Green, The path to 25% silicon solar cell efficiency: history of silicon cell evolution. Prog. Photovolt. Res. Appl. **17**(3), 183–189 (2009)
6. G. Xing et al., Low-temperature solution-processed wavelength-tunable perovskites for lasing. Nat. Mater. **13**(5), 476–480 (2014)
7. M. Cadelano et al., Can trihalide lead perovskites support continuous wave lasing? Adv. Opt. Mater. **3**(11), 1557–1564 (2015)
8. T.C. Sum, N. Mathews, Advancements in perovskite solar cells: photophysics behind the photovoltaics. Energy Environ. Sci. **7**(8), 2518–2534 (2014)
9. W. Ke, M.G. Kanatzidis, Prospects for low-toxicity lead-free perovskite solar cells. Nat. Commun. **10**(1), 965 (2019)
10. M. Becker, T. Klüner, M. Wark, Formation of hybrid ABX_3 perovskite compounds for solar cell application: first-principles calculations of effective ionic radii and determination of tolerance factors. Dalton Trans. **46**(11), 3500–3509 (2017)
11. J.C. Thomas, J.S. Bechtel, A.R. Natarajan, A. Van der Ven, Machine learning the density functional theory potential energy surface for the inorganic halide perovskite $CsPbBr_3$. Phys. Rev. B **100**(13), 134101 (2019)
12. K. Brown, S. Parker, I.R. García, S. Mukhopadhyay, V.G. Sakai, C. Stock, Molecular orientational melting within a lead-halide octahedron framework: the order-disorder transition in $CH_3NH_3PbBr_3$. Phys. Rev. B **96**(17), 174111 (2017)
13. C.C. Stoumpos, C.D. Malliakas, M.G. Kanatzidis, Semiconducting tin and lead iodide perovskites with organic cations: phase transitions, high mobilities, and near-infrared photoluminescent properties. Inorg. Chem. **52**(15), 9019–9038 (2013)
14. R.X. Yang, J.M. Skelton, E.L. Da Silva, J.M. Frost, A. Walsh, Spontaneous octahedral tilting in the cubic inorganic cesium halide perovskites $CsSnX_3$ and $CsPbX_3$ (X= F, Cl, Br, I). J. Phys. Chem. Lett. **8**(19), 4720–4726 (2017)
15. T. Whitcher et al., Dual phases of crystalline and electronic structures in the nanocrystalline perovskite $CsPbBr_3$. NPG Asia Materials **11**(1), 1–12 (2019)
16. N.K. Elumalai, M.A. Mahmud, D. Wang, A. Uddin, perovskite solar cells: progress and advancements. Energies **9**(11), 861 (2016)
17. Y. Jia, R.A. Kerner, A.J. Grede, A.N. Brigeman, B.P. Rand, N.C. Giebink, Diode-pumped organo-lead halide perovskite lasing in a metal-clad distributed feedback resonator. Nano Lett. **16**(7), 4624–4629 (2016)
18. Y. Jia, R.A. Kerner, A.J. Grede, B.P. Rand, N.C. Giebink, Continuous-wave lasing in an organic–inorganic lead halide perovskite semiconductor. Nat. Photonics **11**(12), 784–788 (2017)
19. T. Ungár, *Industrial Applications of X-ray Diffraction*, ed. by F.H. Chung, D.K. Smith (Marcel Dekker, New York, 2000)

20. G.W. Adhyaksa et al., Understanding detrimental and beneficial grain boundary effects in halide perovskites. Adv. Mater. **30**(52), 1804792 (2018)
21. P. Lindley, D. Moss, Elements of X-ray crystallography by LV Azaroff. Acta Crystallogr. Sect. A: Cryst. Phys. Diffr. Theor. Gen. Crystallogr. **26**(6), 701–701 (1970)
22. J. Ding, X. Cheng, L. Jing, T. Zhou, Y. Zhao, S. Du, Polarization-dependent optoelectronic performances in hybrid halide perovskite $MAPbX_3$ (X= Br, Cl) single-crystal photodetectors. ACS Appl. Mater. Interfaces **10**(1), 845–850 (2018)
23. W. Peng et al., Solution-grown monocrystalline hybrid perovskite films for hole-transporter-free solar cells. Adv. Mater. **28**(17), 3383–3390 (2016)
24. F. Zhang et al., Brightly luminescent and color-tunable colloidal $CH_3NH_3PbX_3$ (X= Br, I, Cl) quantum dots: potential alternatives for display technology. ACS Nano **9**(4), 4533–4542 (2015)
25. S.A. Veldhuis et al., Benzyl alcohol-treated $CH_3NH_3PbBr_3$ nanocrystals exhibiting high luminescence, stability, and ultralow amplified spontaneous emission thresholds. Nano Lett. **17**(12), 7424–7432 (2017)
26. L.-C. Chen, K.-L. Lee, C.-Y. Huang, J.-C. Lin, Z.-L. Tseng, Preparation and characteristics of $MAPbBr_3$ perovskite quantum dots on NiO_x film and application for high transparent solar cells. Micromachines **9**(5), 205 (2018)
27. M. Li et al., Amplified spontaneous emission based on 2D Ruddlesden-Popper perovskites. Adv. Funct. Mater. **28**(17), 1707006 (2018). https://doi.org/10.1002/adfm.201707006
28. C.C. Stoumpos et al., Ruddlesden–Popper hybrid lead iodide perovskite 2D homologous semiconductors. Chem. Mater. **28**(8), 2852–2867 (2016)
29. Y. Hua et al., Identification of the band gap energy of two-dimensional $(OA)_2(MA)_{n-1}Pb_nI3_{n+1}$ perovskite with up to 10 layers. J. Phys. Chem. Lett. **10**(22), 7025–7030 (2019)
30. C.M. Raghavan et al., Low-threshold lasing from 2D homologous organic–inorganic hybrid Ruddlesden-Popper perovskite single crystals. Nano Lett. **18**(5), 3221–3228 (2018)
31. I.C. Smith, E.T. Hoke, D. Solis-Ibarra, M.D. McGehee, H.I. Karunadasa, A layered hybrid perovskite solar-cell absorber with enhanced moisture stability. Angew. Chem. **126**(42), 11414–11417 (2014)
32. Y. Chen, Y. Sun, J. Peng, J. Tang, K. Zheng, Z. Liang, 2D Ruddlesden-Popper perovskites for optoelectronics. Adv. Mater. **30**(2), 1703487 (2018)
33. T. Ishihara, J. Takahashi, T. Goto, Exciton state in two-dimensional perovskite semiconductor $(C_{10}H_{21}NH_3)_2PbI_4$. Solid State Commun. **69**(9), 933–936 (1989)
34. T. Ishihara, Optical Properties of Pb-based inorganic-organic perovskites, in *Optical Properties of Low–Dimensional Materials* (World Scientific, 1995), pp. 288–339
35. I. Koutselas, L. Ducasse, G.C. Papavassiliou, Electronic properties of three-and low-dimensional semiconducting materials with Pb halide and Sn halide units. J. Phys.: Condens. Matter **8**(9), 1217 (1996)
36. Y. Lin, Y. Bai, Y. Fang, Q. Wang, Y. Deng, J. Huang, Suppressed ion migration in low-dimensional perovskites. ACS Energy Lett. **2**(7), 1571–1572 (2017)
37. M.R. Leyden, T. Matsushima, C. Qin, S. Ruan, H. Ye, C. Adachi, Amplified spontaneous emission in phenylethylammonium methylammonium lead iodide quasi-2D perovskites. Phys. Chem. Chem. Phys. **20**(22), 15030–15036 (2018)
38. Y.-H. Kim, H. Cho, T.-W. Lee, Metal halide perovskite light emitters. Proc. Natl. Acad. Sci. **113**(42), 11694–11702 (2016)
39. S. Cohen, QLED vs. OLED TV: what's the difference, and why does it matter? Digital Trends. https://www.digitaltrends.com/home-theater/qled-vs-oled-tv/
40. L. Duan et al., Solution processable small molecules for organic light-emitting diodes. J. Mater. Chem. **20**(31), 6392–6407 (2010)
41. S. Reineke et al., White organic light-emitting diodes with fluorescent tube efficiency. Nature **459**(7244), 234–238 (2009)
42. K. Shanmugasundaram, M.S. Subeesh, C.D. Sunesh, Y. Choe, Non-doped deep blue light-emitting electrochemical cells from charged organic small molecules. RSC Adv. **6**(34), 28912–28918 (2016)

43. M.A. Boles, D. Ling, T. Hyeon, D.V. Talapin, The surface science of nanocrystals. Nat. Mater. **15**(2), 141–153 (2016)
44. C. Pal et al., Charge transport in lead sulfide quantum dots/phthalocyanines hybrid nanocomposites. Org. Electron. **44**, 132–143 (2017)
45. C. Dang, A. Nurmikko, Beyond quantum dot LEDs: optical gain and laser action in red, green, and blue colors. MRS Bull. **38**(9), 737–742 (2013)
46. D. Fuhrmann, C. Netzel, U. Rossow, A. Hangleiter, G. Ade, P. Hinze, Optimization scheme for the quantum efficiency of GaInN-based green-light-emitting diodes. Appl. Phys. Lett. **88**(7), 071105 (2006)
47. J. Lingrong et al., GaN-based green laser diodes. J. Semicond. **37**(11), 111001 (2016)
48. J.M. Phillips et al., Research challenges to ultra-efficient inorganic solid-state lighting. Laser Photonics Rev. **1**(4), 307–333 (2007)
49. J. Wu, When group-III nitrides go infrared: New properties and perspectives. J. Appl. Phys. **106**(1), 5 (2009)
50. C.K. Ng, W. Yin, H. Li, J.J. Jasieniak, Scalable synthesis of colloidal $CsPbBr_3$ perovskite nanocrystals with high reaction yields through solvent and ligand engineering. Nanoscale **12**(8), 4859–4867 (2020)
51. G. Nedelcu, L. Protesescu, S. Yakunin, M.I. Bodnarchuk, M.J. Grotevent, M.V. Kovalenko, Fast anion-exchange in highly luminescent nanocrystals of cesium lead halide perovskites ($CsPbX_3$, X= Cl, Br, I). Nano Lett. **15**(8), 5635–5640 (2015)
52. S. Yakunin et al., Low-threshold amplified spontaneous emission and lasing from colloidal nanocrystals of caesium lead halide perovskites. Nat. Commun. **6**(1), 1–9 (2015)
53. B.R. Sutherland, E.H. Sargent, perovskite photonic sources. Nat. Photonics **10**(5), 295 (2016)
54. A. Swarnkar, R. Chulliyil, V.K. Ravi, M. Irfanullah, A. Chowdhury, A. Nag, Colloidal $CsPbBr_3$ perovskite nanocrystals: luminescence beyond traditional quantum dots. Angew. Chem. **127**(51), 15644–15648 (2015)
55. H. Zhu et al., Lead halide perovskite nanowire lasers with low lasing thresholds and high quality factors. Nat. Mater. **14**(6), 636–642 (2015)
56. B.R. Sutherland et al., perovskite thin films via atomic layer deposition. Adv. Mater. **27**(1), 53–58 (2015)
57. Y. Xu et al., Two-photon-pumped perovskite semiconductor nanocrystal lasers. J. Am. Chem. Soc. **138**(11), 3761–3768 (2016)
58. E.E. Hoover, J.A. Squier, Advances in multiphoton microscopy technology. Nat. Photonics **7**(2), 93–101 (2013)
59. Z. Gu et al., Two-photon pumped $CH_3NH_3PbBr_3$ perovskite microwire lasers. Adv. Opt. Mater. **4**(3), 472–479 (2016)
60. Y. Mi, Y. Zhong, Q. Zhang, X. Liu, Continuous-wave pumped perovskite lasers. Adv. Opt. Mater. **7**(17), 1900544 (2019)
61. T.C. Sum, M. Righetto, S.S. Lim, Quo vadis, perovskite emitters? J. Chem. Phys. **152**(13), 130901 (2020)
62. J. Moon et al., Environmentally stable room temperature continuous wave lasing in defect-passivated perovskite (2019). arXiv:1909.10097
63. C. Tian, S. Zhao, W. Zhai, C. Ge, G. Ran, Low-threshold room-temperature continuous-wave optical lasing of single-crystalline perovskite in a distributed reflector microcavity. RSC Adv. **9**(62), 35984–35989 (2019)
64. L. Wang et al., Ultralow-threshold and color-tunable continuous-wave lasing at room-temperature from in situ fabricated perovskite quantum dots. J. Phys. Chem. Lett. **10**(12), 3248–3253 (2019)
65. Y. Wang, X. Li, J. Song, L. Xiao, H. Zeng, H. Sun, All-inorganic colloidal perovskite quantum dots: a new class of lasing materials with favorable characteristics. Adv. Mater. **27**(44), 7101–7108 (2015)
66. G.E. Eperon, E. Jedlicka, D.S. Ginger, Biexciton auger recombination differs in hybrid and inorganic halide perovskite quantum dots. J. Phys. Chem. Lett. **9**(1), 104–109 (2018)

67. F. Staub, U. Rau, T. Kirchartz, Statistics of the auger recombination of electrons and holes via defect levels in the band gap—application to lead-halide perovskites. ACS Omega **3**(7), 8009–8016 (2018)
68. N.S. Makarov, S. Guo, O. Isaienko, W. Liu, I. Robel, V.I. Klimov, Spectral and dynamical properties of single excitons, biexcitons, and trions in cesium–lead-halide perovskite quantum dots. Nano Lett. **16**(4), 2349–2362 (2016)
69. K. Chen, A.J. Barker, F.L. Morgan, J.E. Halpert, J.M. Hodgkiss, Effect of carrier thermalization dynamics on light emission and amplification in organometal halide perovskites. J. Phys. Chem. Lett. **6**(1), 153–158 (2015)
70. J.M. Richter et al., Ultrafast carrier thermalization in lead iodide perovskite probed with two-dimensional electronic spectroscopy. Nat. Commun. **8**(1), 1–7 (2017)
71. Y. Fang, H. Wei, Q. Dong, J. Huang, Quantification of re-absorption and re-emission processes to determine photon recycling efficiency in perovskite single crystals. Nat. Commun. **8**, 14417 (2017)
72. L.M. Pazos-Outón et al., Photon recycling in lead iodide perovskite solar cells. Science **351**(6280), 1430–1433 (2016)
73. P. Geiregat et al., Using bulk-like nanocrystals to probe intrinsic optical gain characteristics of inorganic lead halide perovskites. ACS Nano **12**(10), 10178–10188 (2018)
74. J. Shi et al., Low-threshold stimulated emission of hybrid perovskites at room temperature through defect-mediated bound excitons (2019). arXiv:1902.07371
75. Y. Wang, M. Zhi, Y.-Q. Chang, J.-P. Zhang, Y. Chan, Stable, ultralow threshold amplified spontaneous emission from $CsPbBr_3$ nanoparticles exhibiting trion gain. Nano Lett. **18**(8), 4976–4984 (2018)
76. G. Yumoto et al., Hot biexciton effect on optical gain in $CsPbI_3$ perovskite nanocrystals. J. Phys. Chem. Lett. **9**(9), 2222–2228 (2018)
77. J. Navarro-Arenas, I. Suárez, V.S. Chirvony, A.F. Gualdrón-Reyes, I. Mora-Seró, J. Martínez-Pastor, Single-exciton amplified spontaneous emission in thin films of $CsPbX_3$ (X= Br, I) perovskite nanocrystals. J. Phys. Chem. Lett. **10**(20), 6389–6398 (2019)
78. S. Chen, A. Nurmikko, Excitonic gain and laser emission from mixed-cation halide perovskite thin films. Optica **5**(9), 1141–1149 (2018)
79. M.A. Green, A. Ho-Baillie, perovskite solar cells: the birth of a new era in photovoltaics. ACS Energy Lett. **2**(4), 822–830 (2017)
80. W. Tress et al., Performance of perovskite solar cells under simulated temperature-illumination real-world operating conditions. Nat. Energy **4**(7), 568–574 (2019)
81. Y. Zhang et al., Achieving reproducible and high-efficiency (>21%) perovskite solar cells with a presynthesized $FAPbI_3$ powder. ACS Energy Lett. **5**(2), 360–366 (2019)
82. N.A.N. Ouedraogo et al., Stability of all-inorganic perovskite solar cells. Nano Energy **67**, 104249 (2020)
83. S.F. Leung et al., A self-powered and flexible organometallic halide perovskite photodetector with very high detectivity. Adv. Mater. **30**(8), 1704611 (2018)
84. L. Shen et al., A self-powered, sub-nanosecond-response solution-processed hybrid perovskite photodetector for time-resolved photoluminescence-lifetime detection. Adv. Mater. **28**(48), 10794–10800 (2016)
85. H. Sun, W. Tian, F. Cao, J. Xiong, L. Li, Ultrahigh-performance self-powered flexible double-twisted fibrous broadband perovskite photodetector. Adv. Mater. **30**(21), 1706986 (2018)
86. C. Bao et al., High performance and stable all-inorganic metal halide perovskite-based photodetectors for optical communication applications. Adv. Mater. **30**(38), 1803422 (2018)
87. W. Wu et al., Flexible photodetector arrays based on patterned $CH_3NH_3PbI_{3-x}Cl_x$ perovskite film for real-time photosensing and imaging. Adv. Mater. **31**(3), 1805913 (2019)
88. X.Y. Chin, D. Cortecchia, J. Yin, A. Bruno, C. Soci, Lead iodide perovskite light-emitting field-effect transistor. Nat. Commun. **6**, 7383 (2015)
89. W. Yu et al., Single crystal hybrid perovskite field-effect transistors. Nat. Commun. **9**(1), 1–10 (2018)

90. O.A. Jaramillo-Quintero, R.S. Sanchez, M. Rincon, I. Mora-Sero, Bright visible-infrared light emitting diodes based on hybrid halide perovskite with Spiro-OMeTAD as a hole-injecting layer. J. Phys. Chem. Lett. **6**(10), 1883–1890 (2015)
91. Q. Wang et al., Efficient sky-blue perovskite light-emitting diodes via photoluminescence enhancement. Nat. Commun. **10**(1), 1–8 (2019)
92. Z.-K. Tan et al., Bright light-emitting diodes based on organometal halide perovskite. Nat. Nanotechnol. **9**(9), 687–692 (2014)
93. S. Pathak et al., perovskite crystals for tunable white light emission. Chem. Mater. **27**(23), 8066–8075 (2015)
94. H. Huang et al., Water resistant $CsPbX_3$ nanocrystals coated with polyhedral oligomeric silsesquioxane and their use as solid state luminophores in all-perovskite white light-emitting devices. Chem. Sci. **7**(9), 5699–5703 (2016)
95. J.R. Harwell, G.L. Whitworth, G.A. Turnbull, I.D.W. Samuel, Green perovskite distributed feedback lasers. Sci. Rep. **7**(1), 1–8 (2017)
96. N. Pourdavoud et al., Photonic nanostructures patterned by thermal nanoimprint directly into organo-metal halide perovskites. Adv. Mater. **29**(12), 1605003 (2017)
97. M. Saliba et al., Structured organic–inorganic perovskite toward a distributed feedback laser. Adv. Mater. **28**(5), 923–929 (2016)
98. F. Mathies, P. Brenner, G. Hernandez-Sosa, I.A. Howard, U.W. Paetzold, U. Lemmer, Inkjet-printed perovskite distributed feedback lasers. Opt. Express **26**(2), A144–A152 (2018)
99. Z. Wei, H. Chen, K. Yan, S. Yang, Inkjet printing and instant chemical transformation of a $CH_3NH_3PbI_3$/nanocarbon electrode and interface for planar perovskite solar cells. Angew. Chem. Int. Ed. **53**(48), 13239–13243 (2014)
100. L. Shi et al., In situ inkjet printing strategy for fabricating perovskite quantum dot patterns. Adv. Func. Mater. **29**(37), 1903648 (2019)
101. W. Nie et al., High-efficiency solution-processed perovskite solar cells with millimeter-scale grains. Science **347**(6221), 522–525 (2015)
102. D. Shi et al., Low trap-state density and long carrier diffusion in organolead trihalide perovskite single crystals. Science **347**(6221), 519–522 (2015)
103. V. Adinolfi et al., The in-gap electronic state spectrum of methylammonium lead iodide single-crystal perovskites. Adv. Mater. **28**(17), 3406–3410 (2016)
104. Q. Dong et al., Electron-hole diffusion lengths >175 μm in solution-grown $CH_3NH_3PbI_3$ single crystals. Science **347**(6225), 967–970 (2015)
105. B. Wu et al., Discerning the surface and bulk recombination kinetics of organic–inorganic halide perovskite single crystals. Adv. Energy Mater. **6**(14), 1600551 (2016)
106. D. Yang et al., Amplified spontaneous emission from organic–inorganic hybrid lead iodide perovskite single crystals under direct multiphoton excitation. Adv. Opt. Mater. **4**(7), 1053–1059 (2016)
107. C. Zhao et al., Stable two-photon pumped amplified spontaneous emission from millimeter-sized $CsPbBr_3$ single crystals. J. Phys. Chem. Lett. **10**(10), 2357–2362 (2019)
108. A.O. Murzin et al., Amplified spontaneous emission and random lasing in $MAPbBr_3$ halide perovskite single crystals. Adv. Opt. Mater. 2000690 (2020)
109. B. Li et al., Dynamic growth of pinhole-free conformal $CH_3NH_3PbI_3$ film for perovskite solar cells. ACS Appl. Mater. Interfaces **8**(7), 4684–4690 (2016)
110. S.S. Lim et al., Modulating carrier dynamics through perovskite film engineering. Phys. Chem. Chem. Phys. **18**(39), 27119–27123 (2016)
111. C. Bi et al., Understanding the formation and evolution of interdiffusion grown organolead halide perovskite thin films by thermal annealing. J. Mater. Chem. A **2**(43), 18508–18514 (2014)
112. A. Dualeh, N. Tétreault, T. Moehl, P. Gao, M.K. Nazeeruddin, M. Grätzel, Effect of annealing temperature on film morphology of organic–inorganic hybrid pervoskite solid-state solar cells. Adv. Func. Mater. **24**(21), 3250–3258 (2014)
113. N. Yantara et al., Inorganic halide perovskites for efficient light-emitting diodes. J. Phys. Chem. Lett. **6**(21), 4360–4364 (2015)

114. G.E. Eperon, V.M. Burlakov, P. Docampo, A. Goriely, H.J. Snaith, Morphological control for high performance, solution-processed planar heterojunction perovskite solar cells. Adv. Func. Mater. **24**(1), 151–157 (2014)
115. L. Huang et al., $CH_3NH_3PbI_{3-x}Cl_x$ films with coverage approaching 100% and with highly oriented crystal domains for reproducible and efficient planar heterojunction perovskite solar cells. Phys. Chem. Chem. Phys. **17**(34), 22015–22022 (2015)
116. H. Cho et al., Overcoming the electroluminescence efficiency limitations of perovskite light-emitting diodes. Science **350**(6265), 1222–1225 (2015). https://doi.org/10.1126/science.aad1818
117. J.J. Yoo et al., An interface stabilized perovskite solar cell with high stabilized efficiency and low voltage loss. Energy Environ. Sci. **12**(7), 2192–2199 (2019)
118. D.-Y. Son et al., Self-formed grain boundary healing layer for highly efficient $CH_3NH_3PbI_3$ perovskite solar cells. Nat. Energy **1**(7), 1–8 (2016)
119. D. Wang, M. Wright, N.K. Elumalai, A. Uddin, Stability of perovskite solar cells. Sol. Energy Mater. Sol. Cells **147**, 255–275 (2016)
120. Y. Shao et al., Grain boundary dominated ion migration in polycrystalline organic–inorganic halide perovskite films. Energy Environ. Sci. **9**(5), 1752–1759 (2016)
121. Y. Yuan, J. Huang, Ion migration in organometal trihalide perovskite and its impact on photovoltaic efficiency and stability. Acc. Chem. Res. **49**(2), 286–293 (2016)
122. N.J. Jeon, J.H. Noh, Y.C. Kim, W.S. Yang, S. Ryu, S.I. Seok, Solvent engineering for high-performance inorganic–organic hybrid perovskite solar cells. Nat. Mater. **13**(9), 897–903 (2014)
123. E. Lafalce, C. Zhang, Y. Zhai, D. Sun, Z. Vardeny, Enhanced emissive and lasing characteristics of nano-crystalline MAPbBr3 films grown via anti-solvent precipitation. J. Appl. Phys. **120**(14), 143101 (2016)
124. A. Gharajeh et al., Amplified spontaneous emission in nanoimprinted perovskite nanograting metasurface, in *2017 IEEE 17th International Conference on Nanotechnology (IEEE-NANO)* (IEEE, 2017), pp. 534–536
125. F. Deschler et al., High photoluminescence efficiency and optically pumped lasing in solution-processed mixed halide perovskite semiconductors. J. Phys. Chem. Lett. **5**(8), 1421–1426 (2014)
126. N. Arora et al., Photovoltaic and amplified spontaneous emission studies of high-quality formamidinium lead bromide perovskite films. Adv. Func. Mater. **26**(17), 2846–2854 (2016)
127. L. Zhang et al., One-step co-evaporation of all-inorganic perovskite thin films with room-temperature ultralow amplified spontaneous emission threshold and air stability. ACS Appl. Mater. Interfaces **10**(47), 40661–40671 (2018)
128. M.L. De Giorgi, A. Perulli, N. Yantara, P.P. Boix, M. Anni, Amplified spontaneous emission properties of solution processed $CsPbBr_3$ perovskite thin films. J. Phys. Chem. C **121**(27), 14772–14778 (2017)
129. N. Pourdavoud et al., Room-temperature stimulated emission and lasing in recrystallized cesium lead bromide perovskite thin films. Adv. Mater. **31**(39), 1903717 (2019)
130. S.D. Stranks et al., Enhanced amplified spontaneous emission in perovskites using a flexible cholesteric liquid crystal reflector. Nano Lett. **15**(8), 4935–4941 (2015)
131. J. Li, J. Si, L. Gan, Y. Liu, Z. Ye, H. He, Simple approach to improving the amplified spontaneous emission properties of perovskite films. ACS Appl. Mater. Interfaces **8**(48), 32978–32983 (2016)
132. L. Qin et al., Enhanced amplified spontaneous emission from morphology-controlled organic–inorganic halide perovskite films. RSC Adv. **5**(125), 103674–103679 (2015)
133. H. Ren et al., Efficient and stable Ruddlesden-Popper perovskite solar cell with tailored interlayer molecular interaction. Nat. Photonics **14**(3), 154–163 (2020)
134. C. Fang et al., High-performance photodetectors based on lead-free 2D Ruddlesden-Popper perovskite/MoS_2 heterostructures. ACS Appl. Mater. Interfaces **11**(8), 8419–8427 (2019)
135. R. Dong et al., Novel series of quasi-2D Ruddlesden-Popper perovskites based on short-chained spacer cation for enhanced photodetection. ACS Appl. Mater. Interfaces **10**(22), 19019–19026 (2018)

136. T. Matsushima et al., Solution-processed organic-inorganic perovskite field-effect transistors with high hole mobilities. Adv. Mater. **28**(46), 10275–10281 (2016)
137. X. Yang et al., Efficient green light-emitting diodes based on quasi-two-dimensional composition and phase engineered perovskite with surface passivation. Nat. Commun. **9**(1), 1–8 (2018)
138. L. Mao, Y. Wu, C.C. Stoumpos, M.R. Wasielewski, M.G. Kanatzidis, White-light emission and structural distortion in new corrugated two-dimensional lead bromide perovskites. J. Am. Chem. Soc. **139**(14), 5210–5215 (2017)
139. K. Thirumal et al., Morphology-independent stable white-light emission from self-assembled two-dimensional perovskites driven by strong exciton–phonon coupling to the organic framework. Chem. Mater. **29**(9), 3947–3953 (2017)
140. W. Deng, X. Jin, Y. Lv, X. Zhang, X. Zhang, J. Jie, 2D Ruddlesden-popper perovskite nanoplate based deep-blue light-emitting diodes for light communication. Adv. Func. Mater. **29**(40), 1903861 (2019)
141. P. Vashishtha, M. Ng, S.B. Shivarudraiah, J.E. Halpert, High efficiency blue and green light-emitting diodes using Ruddlesden-Popper inorganic mixed halide perovskites with butylammonium interlayers. Chem. Mater. **31**(1), 83–89 (2018)
142. T. Kondo, T. Azuma, T. Yuasa, R. Ito, Biexciton lasing in the layered perovskite-type material $(C_6H_{13}NH_3)_2PbI_4$. Solid State Commun. **105**(4), 253–255 (1998)
143. W. Zhai et al., Optically pumped lasing of segregated quasi-2D perovskite microcrystals in vertical microcavity at room temperature. Appl. Phys. Lett. **114**(13), 131107 (2019)
144. M. Xia et al., Two-dimensional perovskites as sensitive strain sensors. J. Mater. Chem. C **8**(11), 3814–3820 (2020)
145. A. Feng et al., Shape control of metal halide perovskite single crystals: from bulk to nanoscale. Chem. Mater. **32**(18), 7602–7617 (2020)
146. C. Huang et al., Up-conversion perovskite nanolaser with single mode and low threshold. J. Phys. Chem. C **121**(18), 10071–10077 (2017)
147. M. Li et al., Enhanced exciton and photon confinement in Ruddlesden-Popper perovskite microplatelets for highly stable low-threshold polarized lasing. Adv. Mater. **30**(23), 1707235 (2018)
148. B. Yang et al., Low threshold two-photon-pumped amplified spontaneous emission in $CH_3NH_3PbBr_3$ microdisks. ACS Appl. Mater. Interfaces **8**(30), 19587–19592 (2016)
149. W. Zhang et al., Controlling the cavity structures of two-photon-pumped perovskite microlasers. Adv. Mater. **28**(21), 4040–4046 (2016)
150. M.K. Hossain et al., Controllable optical emission wavelength in all-inorganic halide perovskite alloy microplates grown by two-step chemical vapor deposition. Nano Res. **13**(11), 2939–2949 (2020)
151. Q. Zhang, S.T. Ha, X. Liu, T.C. Sum, Q. Xiong, Room-temperature near-infrared high-Q perovskite whispering-gallery planar nanolasers. Nano Lett. **14**(10), 5995–6001. https://doi.org/10.1021/nl503057g
152. Q. Zhang, R. Su, X. Liu, J. Xing, T.C. Sum, Q. Xiong, High-quality whispering-gallery-mode lasing from cesium lead halide perovskite nanoplatelets. Adv. Func. Mater. **26**(34), 6238–6245 (2016)
153. P. Perumal, C.-S. Wang, K.M. Boopathi, G. Haider, W.-C. Liao, Y.-F. Chen, Whispering gallery mode lasing from self-assembled hexagonal perovskite single crystals and porous thin films decorated by dielectric spherical resonators. ACS Photonics **4**(1), 146–155 (2017)
154. A. Zhizhchenko et al., Single-mode lasing from imprinted halide-perovskite microdisks. ACS Nano **13**(4), 4140–4147 (2019)
155. J. Ward, O. Benson, WGM microresonators: sensing, lasing and fundamental optics with microspheres. Laser Photonics Rev. **5**(4), 553–570 (2011)
156. G. Li et al., Record-low-threshold lasers based on atomically smooth triangular nanoplatelet perovskite. Adv. Func. Mater. **29**(2), 1805553 (2019)
157. R. Su et al., Room-temperature polariton lasing in all-inorganic perovskite nanoplatelets. Nano Lett. **17**(6), 3982–3988 (2017)

158. X. Liu et al., Periodic organic-inorganic halide perovskite microplatelet arrays on silicon substrates for room-temperature lasing. Adv. Sci. **3**(11), 1600137 (2016)
159. J. Feng et al., "Liquid knife" to fabricate patterning single-crystalline perovskite microplates toward high-performance laser arrays. Adv. Mater. **28**(19), 3732–3741 (2016)
160. K. Wang, Z. Gu, S. Liu, J. Li, S. Xiao, Q. Song, Formation of single-mode laser in transverse plane of perovskite microwire via micromanipulation. Opt. Lett. **41**(3), 555–558 (2016)
161. S.A. Veldhuis et al., perovskite materials for light-emitting diodes and lasers. Adv. Mater. **28**(32), 6804–6834 (2016)
162. K. Wang et al., High-density and uniform lead halide perovskite nanolaser array on silicon. J. Phys. Chem. Lett. **7**(13), 2549–2555 (2016)
163. Y. Fu et al., Nanowire lasers of formamidinium lead halide perovskites and their stabilized alloys with improved stability. Nano Lett. **16**(2), 1000–1008 (2016)
164. J. Xing et al., Vapor phase synthesis of organometal halide perovskite nanowires for tunable room-temperature nanolasers. Nano Lett. **15**(7), 4571–4577 (2015)
165. E.T. Hoke, D.J. Slotcavage, E.R. Dohner, A.R. Bowring, H.I. Karunadasa, M.D. McGehee, Reversible photo-induced trap formation in mixed-halide hybrid perovskites for photovoltaics. Chem. Sci. **6**(1), 613–617 (2015)
166. S. Zhang et al., Strong exciton–photon coupling in hybrid inorganic–organic perovskite micro/nanowires. Adv. Opt. Mater. **6**(2), 1701032 (2018)
167. X. Wang et al., High-quality in-plane aligned $CsPbX_3$ perovskite nanowire lasers with composition-dependent strong exciton–photon coupling. ACS Nano **12**(6), 6170–6178 (2018)
168. K. Park et al., Light–matter interactions in cesium lead halide perovskite nanowire lasers. J. Phys. Chem. Lett. **7**(18), 3703–3710 (2016)
169. W. Du et al., Strong exciton–photon coupling and lasing behavior in all-inorganic $CsPbBr_3$ micro/nanowire Fabry-Pérot cavity. ACS Photonics **5**(5), 2051–2059 (2018)
170. Q. Shang et al., Enhanced optical absorption and slowed light of reduced-dimensional $CsPbBr_3$ nanowire crystal by exciton-polariton. Nano Lett. **20**(2), 1023–1032 (2020)
171. A.P. Schlaus et al., How lasing happens in $CsPbBr_3$ perovskite nanowires. Nat. Commun. **10**(1), 1–8 (2019)
172. F. Li et al., Controlled fabrication, lasing behavior and excitonic recombination dynamics in single crystal $CH_3NH_3PbBr_3$ perovskite cuboids. Sci. Bull. **64**(10), 698–704 (2019)
173. Z.-Y. Zhang et al., Size-dependent one-photon-and two-photon-pumped amplified spontaneous emission from organometal halide $CH_3NH_3PbBr_3$ perovskite cubic microcrystals. Phys. Chem. Chem. Phys. **19**(3), 2217–2224 (2017)
174. Z. Hu et al., Robust cesium lead halide perovskite microcubes for frequency upconversion lasing. Adv. Opt. Mater. **5**(22), 1700419 (2017)
175. D. Yang, S. Chu, Y. Wang, C.K. Siu, S. Pan, S.F. Yu, Frequency upconverted amplified spontaneous emission and lasing from inorganic perovskite under simultaneous six-photon absorption. Opt. Lett. **43**(9), 2066–2069 (2018)
176. B. Zhou et al., Single-mode lasing and 3D confinement from perovskite micro-cubic cavity. J. Mater. Chem. C **6**(43), 11740–11748 (2018)
177. A. Chiasera et al., Spherical whispering-gallery-mode microresonators. Laser Photonics Rev. **4**(3), 457–482 (2010)
178. F. Chen et al., Detachable surface plasmon substrate to enhance $CH_3NH_3PbBr_3$ lasing. Opt. Commun. **452**, 400–404 (2019)
179. Q.A. Akkerman et al., Tuning the optical properties of cesium lead halide perovskite nanocrystals by anion exchange reactions. J. Am. Chem. Soc. **137**(32), 10276–10281 (2015)
180. C. Guhrenz, A. Benad, C. Ziegler, D. Haubold, N. Gaponik, A. Eychmüller, Solid-state anion exchange reactions for color tuning of $CsPbX_3$ perovskite nanocrystals. Chem. Mater. **28**(24), 9033–9040 (2016)
181. G. Li, J.Y.-L. Ho, M. Wong, H.S. Kwok, Reversible anion exchange reaction in solid halide perovskites and its implication in photovoltaics. J. Phys. Chem. C **119**(48), 26883–26888 (2015)

182. D. Parobek, Y. Dong, T. Qiao, D. Rossi, D.H. Son, Photoinduced anion exchange in cesium lead halide perovskite nanocrystals. J. Am. Chem. Soc. **139**(12), 4358–4361 (2017)
183. L. Protesescu et al., Nanocrystals of cesium lead halide perovskites ($CsPbX_3$, X= Cl, Br, and I): novel optoelectronic materials showing bright emission with wide color gamut. Nano Lett. **15**(6), 3692–3696 (2015)
184. S. Wei, Y. Yang, X. Kang, L. Wang, L. Huang, D. Pan, Room-temperature and gram-scale synthesis of $CsPbX_3$ (X= Cl, Br, I) perovskite nanocrystals with 50–85% photoluminescence quantum yields. Chem. Commun. **52**(45), 7265–7268 (2016)
185. X. Li et al., Healing all-inorganic perovskite films via recyclable dissolution–recyrstallization for compact and smooth carrier channels of optoelectronic devices with high stability. Adv. Func. Mater. **26**(32), 5903–5912 (2016)
186. F. Zhang et al., Colloidal synthesis of air-stable $CH_3NH_3PbI_3$ quantum dots by gaining chemical insight into the solvent effects. Chem. Mater. **29**(8), 3793–3799 (2017)
187. C.J. Chang-Hasnain, Tunable VCSEL. IEEE J. Sel. Top. Quantum Electron. **6**(6), 978–987 (2000)
188. M. Imran et al., Benzoyl halides as alternative precursors for the colloidal synthesis of lead-based halide perovskite nanocrystals. J. Am. Chem. Soc. **140**(7), 2656–2664 (2018)
189. S. Chang, Z. Bai, H. Zhong, In situ fabricated perovskite nanocrystals: a revolution in optical materials. Adv. Opt. Mater. **6**(18), 1800380 (2018)
190. M.-H. Park et al., Unravelling additive-based nanocrystal pinning for high efficiency organic-inorganic halide perovskite light-emitting diodes. Nano Energy **42**, 157–165 (2017)
191. J.-W. Lee et al., In-situ formed type I nanocrystalline perovskite film for highly efficient light-emitting diode. ACS Nano **11**(3), 3311–3319 (2017)
192. L. Zhao et al., In situ preparation of metal halide perovskite nanocrystal thin films for improved light-emitting devices. ACS Nano **11**(4), 3957–3964 (2017)
193. D. Di et al., Size-dependent photon emission from organometal halide perovskite nanocrystals embedded in an organic matrix. J. Phys. Chem. Lett. **6**(3), 446–450 (2015)
194. Q. Zhou, Z. Bai, W. G. Lu, Y. Wang, B. Zou, H. Zhong, In situ fabrication of halide perovskite nanocrystal-embedded polymer composite films with enhanced photoluminescence for display backlights. Adv. Mater. **28**(41), 9163–9168 (2016)
195. Y. Xin, H. Zhao, J. Zhang, Highly stable and luminescent perovskite–polymer composites from a convenient and universal strategy. ACS Appl. Mater. Interfaces **10**(5), 4971–4980 (2018)
196. L.N. Quan et al., Highly emissive green perovskite nanocrystals in a solid state crystalline matrix. Adv. Mater. **29**(21), 1605945 (2017)
197. X. Chen et al., Centimeter-sized Cs_4PbBr_6 crystals with embedded $CsPbBr_3$ nanocrystals showing superior photoluminescence: nonstoichiometry induced transformation and light-emitting applications. Adv. Func. Mater. **28**(16), 1706567 (2018)
198. Y. Wang, X. Li, X. Zhao, L. Xiao, H. Zeng, H. Sun, Nonlinear absorption and low-threshold multiphoton pumped stimulated emission from all-inorganic perovskite nanocrystals. Nano Lett. **16**(1), 448–453 (2016)
199. F. Todescato et al., Soft-lithographed up-converted distributed feedback visible lasers based on CdSe–CdZnS–ZnS quantum dots. Adv. Func. Mater. **22**(2), 337–344 (2012)
200. J.J. Jasieniak et al., Highly efficient amplified stimulated emission from CdSe-CdS-ZnS quantum dot doped waveguides with two-photon infrared optical pumping. Adv. Mater. **20**(1), 69–73 (2008)
201. Z. Liu et al., Two-photon pumped amplified spontaneous emission and lasing from formamidinium lead bromine nanocrystals. ACS Photonics **6**(12), 3150–3158 (2019)
202. M. Sheik-Bahae, A.A. Said, T.-H. Wei, D.J. Hagan, E.W. Van Stryland, Sensitive measurement of optical nonlinearities using a single beam. IEEE J. Quantum Electron. **26**(4), 760–769 (1990)

Chapter 3
Optical Gain Mechanisms and Fabrication of Perovskite Lasers

3.1 Optical Gain Mechanisms in Halide Perovskites

In this section, we shall focus on the photophysics of optical gain in Perovskites. Specifically, we are interested in using optical spectroscopy to uncover the underlying carrier dynamics leading to the ASE or lasing in Perovskites. Thus, we begin by introducing basic optical spectroscopic tools used in related studies in Sect. 3.1.1 while Sects. 3.1.2, 3.1.3, 3.1.4, 3.1.5 and 3.1.6 discuss spectroscopic results providing evidence for various optical gain mechanisms.

3.1.1 Spectroscopic Tools for Studying Optical Gain

A clear understanding of the underlying carrier dynamics in Perovskite gain materials will provide valuable insights on device integration and optimisation. Since carrier dynamics in materials span over wide (sub-pico, pico-, nano- and even micro-second) timescales, excitation sources must be temporally faster than the specific process under probe. As such, time-sensitive tools such as ultrafast optical spectroscopy (UOS) are often relied on to conduct such studies. Essentially, laser pulses with femtosecond time-widths and high peak instantaneous intensities serve as excitation sources, of which one could perform both time-independent (Steady-state) and time-resolved measurements. Usually, such studies begin with steady-state measurements aiming to verify the existence of optical gain and its gain coefficients in Perovskite materials, followed by time-resolved measurements to explain the steady-state observations. In steady-state measurements, one would obtain the time-averaged emission spectra of a sample, collected by either a charge-coupled device (CCD) or photon-multiplier tube (PMT) over a specified time-window. Fluence-dependent evolution of PL spectral profiles using regular PL or VSL setups are examples of steady-state measurements, as shown in Fig. 3.1.

Y. K. E. Tay et al., *Halide Perovskite Lasers*, Nanoscience and Nanotechnology,
https://doi.org/10.1007/978-981-16-7973-5_3

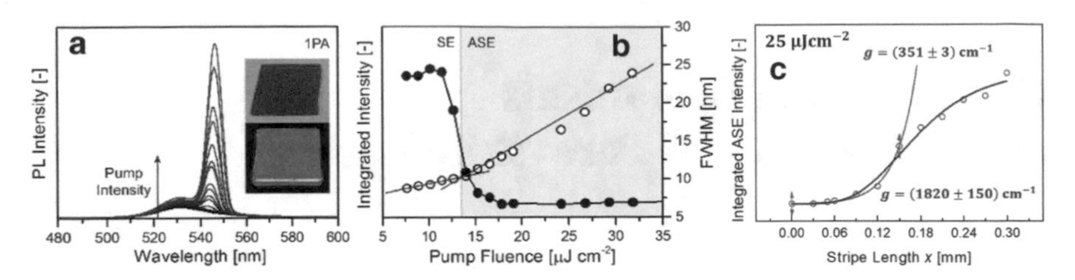

Fig. 3.1 Steady-state characterisation to verify ASE in BnOH-treated $CH_3NH_3PbBr_3$ PNC films [1]. **a** Fluence-dependent evolution of PL spectra, from broad spontaneous emission (SE) band to a relatively red-shifted ASE band. **b** Non-linear increase in spectrally integrated intensities (hollow circles) and FWHM narrowing (solid circles) imply a threshold feature that describes the onset of ASE. **c** Stripe length dependent VSL-PL spectra pumped at fixed 25 μJcm^{-2} evaluates the PNC film's net gain coefficient

As shown in Fig. 3.1b, the non-linear growth in the spectrally integrated intensity followed by a simultaneous FWHM narrowing are preliminary evidence of ASE in Perovskite gain materials. As such, the "kink" where integrated intensity increases super-linearly and a sudden rapid decrease in its FWHM is the ASE pumping threshold [1]. The reason for ASE manifesting in the red-shifted lower energy side depends on the optical gain mechanism. This will be elaborated in Sects. 3.1.2, 3.1.3, 3.1.4 and 3.1.5. Next, as discussed in Sect. 1.2.3, VSL measurements of $CH_3NH_3PbBr_3$ PNC films under a fixed above threshold pumping 25 μJcm^{-2} reveal a general gain coefficient of g $\sim$ 1800 cm^{-2} with an unsaturated gain coefficient of g $\sim$ 350 cm^{-2}. Although not included in the illustration of Fig. 3.1, PL measurements conducted using microscopes called $\mu - PL$ are often used to directly observe lasing emissions in fabricated or natural Perovskite microcavities, as seen in Sect. 2.2.

Moving on, the next set of studies are time-resolved measurements which include time-resolved photoluminescence (TR-PL) and pump-probe (transient absorption, TA) techniques. In most labs, TR-PL measurements are simply conducted by redirecting the Perovskite's output emission into either a streak camera or time-correlated single photon counting (TCSPC) system. Figure 3.2 shows a complete set of TR-PL data of the same $CH_3NH_3PbBr_3$ PNC films [1] collected by a streak camera with resolution down to ~10 ps.

Figure 3.2a shows a spectrographic map, where both time (Fig. 3.2b) and instantaneous steady-state PL (Fig. 3.2c) data could be extracted. With reference to both Fig. 3.2a, c, the longer-lived component is assigned to the broad spontaneous emission driven by excitonic recombination in $CH_3NH_3PbBr_3$ PNCs, while the extremely short-lived red-shifted component corresponds to the narrowband ASE lifetime, assigned to biexcitonic recombination mechanisms [1]. By temporally resolving both components' kinetics as shown in Fig. 3.2b, we obtain information on their processes' lifetimes relative to each other. The kinetics are typically fitted using the multiexponential carrier kinetics function modelled using carrier rate-equations:

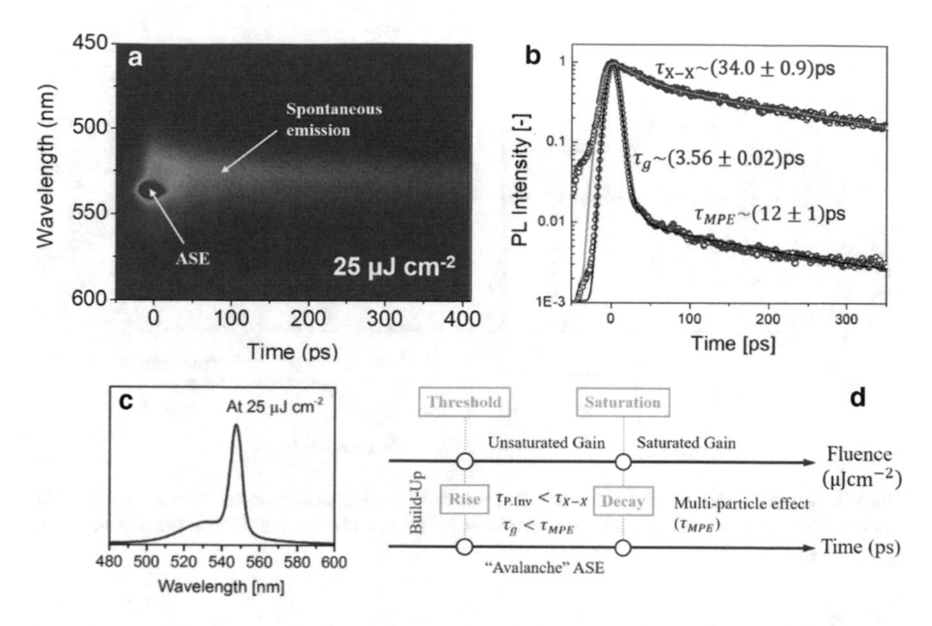

Fig. 3.2 Time-resolved PL (TR-PL) measurement of processes occurring in $CH_3NH_3PbBr_3$ PNCs [1]. **a** The spectrograph containing information of both **b** PL kinetics and **c** instantaneous PL spectra. **d** A timeline of processes both contributing and hindering optical build up and avalanched stimulated emission. The top fluence line helps to capture processes unique to different pumping regimes

$$y(t) = \sum_j A_j e^{-t/\tau_j}\left(1 + \mathrm{erf}\left(-\frac{t - s^2}{\tau_j s}\right)\right) \tag{3.1}$$

where A_j and τ_j are the process' characteristic amplitude and lifetime, respectively and s found in the error-function is the streak camera's response time that accounts for the detected signal rise time. Assignment of fitted lifetimes for each kinetic decay purely from TR-PL data can be an extremely difficult task and would usually require spectral deconvolution of its steady-state PL spectrum in order to draw some conclusions [2]. An in-depth discussion of biexcitonic gain and lifetime assignment to involving processes with respect to Fig. 3.2b, d will be re-visited in Sect. 3.1.3. The time resolution of a streak camera system can be up to several ps, while the TCSPC system is around tens of ps. To probe the fast processes like optical build up and stimulated emission decay times of Perovskite media, femtosecond fluorescence upconversion spectroscopy [3] a gated PL technique, extends the time resolution to tens of femtoseconds.

Pump-probe spectroscopy (also known as transient absorption spectroscopy) is another time-resolved technique that is commonly used to supplement TR-PL data. This is because, the pump-probe measurement allows time resolution down to its

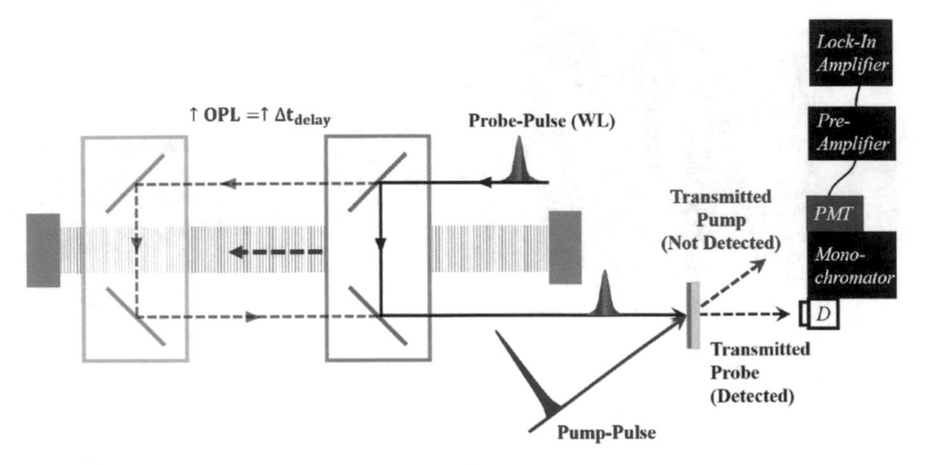

Fig. 3.3 A typical pump-probe experimental setup consists of spatially and temporally overlapping pump and probe pulse trains. Chirp correction at early times is needed for the broad band white light probe

excitation pulse-widths. Hence, pump-probe measurements can resolve and accurately detect both optical gain build up and decay processes. As shown in Fig. 3.3, the superior time resolution is afforded by the detection of pump-induced probe change signals.

In a typical pump-probe experiment in transmission detection mode, the probe pulse is aligned to directly enter the detector while the pump pulse, it is aligned away from the detector. In this way, the detector picks up signals arising from pump-induced probe transmittivity changes within the Perovskite material. A common configuration is one where the probe is a white-light continuum, where its time-delay relative to the pump excitation event is afforded by a controllable mechanical delay stage. Note that the detected probe signal at "time-zero" can be optimised maximally when there is perfect spatial and temporal overlap between both pulses. For the accuracy of experiment, it is important to consider several factors. Firstly, to capture the pump-induced probe changes, the pump's spot size should be bigger than the probe's spot size. Secondly, the fluence of the probe should be significantly lower than the pump's fluence. Lastly, a lock-in amplifier will help improve the signal-to-noise ratio (SNR). Optical choppers with a prime number chopping frequency (e.g., ~83 Hz) are typically placed in the path of the pump beam to isolate electrical noises and its harmonics (e.g., ceiling lamps). For a rotating disc chopper, this also effectively reduces the pump pulse train's repetition rate by a factor of half. A pre-amplifier system can also be added before the lock-in amplifier to filter off unwanted frequencies and to amplify the probe detection along the chopping frequency. Like TR-PL, the pump-probe experiment generates a spectrograph detailing information on both the TA spectra and kinetics of the Perovskite sample. TA spectra provide information on spectral bands where carrier dynamics occur. On the other hand, the TA kinetics as a function of probe delay time tracks the interplay of the various processes and their lifetimes in each spectral band.

Generally, pump-probe experiments done in transmission mode yield the differential transmission $\left(\frac{\Delta T}{T_0}\right)$ probe signals which is defined as:

$$\frac{\Delta T}{T_0} = \frac{T_p - T_0}{T_0} \tag{3.2}$$

where T_p and T_0 denotes the probe signals detected when pump pulses are turned ON and OFF, respectively. Since optical gain is thought to be negative absorption, evidence of light amplification in a material is best interpreted when converting $\left(\frac{\Delta T}{T_0}\right)$ to $\left(\frac{\Delta A}{A_0}\right)$ and has been well-established for chalcogenide CdSe and CdS QDs [4, 5]. In using the Lambert-Beers' law, we obtain the following conversion from $\left(\frac{\Delta T}{T_0}\right)$ to the probes' change in absorbance (ΔA):

$$(\Delta A) = -\log_{10}\left(1 + \frac{\Delta T}{T_0}\right) \tag{3.3}$$

Next, the differential change in probe absorbance $\left(\frac{\Delta A}{A_0}\right)$ can be acquired by simply dividing Eq. (3.3) with the linear absorbance A_0 from the material's absorption spectrum. The onset of optical gain in terms of $\left(\frac{\Delta A}{A_0}\right)$ occurs when the probe picks up a negative signal, such that:

$$A_p = \Delta A + A_0 \leq 0 \tag{3.4}$$

By manipulating Eq. (3.4), we arrive at the threshold condition of:

$$\frac{\Delta A}{A_0} \leq -1 \tag{3.5}$$

In a typical transient kinetics curve, the gain lifetime corresponds to the duration of which the condition in Eq. (3.5) is satisfied and was used as a benchmark for reporting gain lifetimes in CdSe core–shell QDs [5] and $CsPbBr_3$ PNCs with trionic gain mechanisms, as shown in Fig. 3.4 [6].

Two other time-resolved spectroscopic methods occasionally used for studying optical gain in Perovskites are (I) Optical Kerr Gating (OKG) and 2-Dimensional Electron Spectroscopy (2DES). The OKG technique was originally introduced by Duguay and Hansen et al. in 1969 [7] and used by Chen et al. in 2014 to study the effects of carrier thermalisations in free-carrier optical gain in $CH_3NH_3PbI_3$ PTFs [8]. 2DES was also used in optical gain studies in PNCs over conventional TR-PL and pump-probe approaches, as it could resolve ensemble-size induced inhomogeneous broadening [9]. Wei et al. was the first to provide insights of biexcitonic

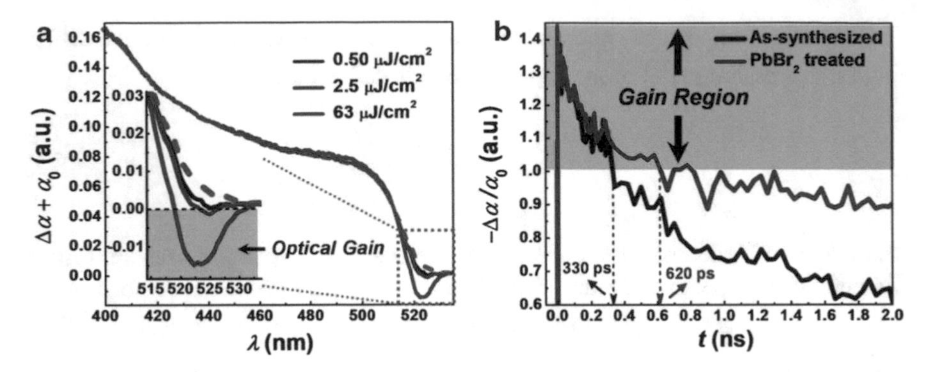

Fig. 3.4 Onset of trionic gain (blue-shaded) in $PbBr_2$ treated $CsPbBr_3$ PNC films, **a** which occurs along the 520–530 nm transition line in the TA spectra and **b** an ASE lifetime of ~620 ps from the TA kinetics [6]

gain in $CsPbBr_3$ PNC films using the 2DES approach and will be discussed more in Sect. 3.1.3. However, we will not go into the experimental details of both techniques.

3.1.2 Free-Carrier Optical Gain

Free-carrier optical gain refers to the population inversion of free-electron carriers in the conduction band (CB) edge relative to its unexcited electrons in valence band (VB). As early as 2014, Chen et al. employed the OKG technique in order to study the role of carrier thermalisation in free-carrier optical gain in $CH_3NH_3PbI_{3-x}Cl_x$ films [8]. Figure 3.5a shows that the ASE onset time and carrier thermalisation time are closely related at ~8 ps and ~10 ps respectively. In addition, Fig. 3.5b showed that the ASE signature occurs along the same peak position as its spontaneous emission. Together, these results suggest that the avalanche stimulated emission occurs along the bandedge, where free-electrons accumulate in the conduction bandedge (CB) during optical build up. Supposing that a highly energetic pump exceeding the bandgap energy is used, this means the free-carrier optical gain relies heavily on hot-carrier thermalisation to facilitate the early time optical build up. For this reason, free-carrier optical gain is often termed as a "thermalisation-limited" mechanism. Interestingly, partial recovery of the ASE can be achieved after the decay of the initial ASE event, due to continuous hot carrier relaxation that helps to replenish free-electrons in CB edge, as shown by the later time "shoulder" in Fig. 3.5c [8]. A schematic of the free-carrier optical gain mechanism is shown in Fig. 3.6, showing that its "thermalisation-limited" trait makes the free-carrier optical gain akin to a three-level system, CB edge that acts as the metastable state, where effective accumulation occurs up till its bandedge radiative recombination lifetime τ_r.

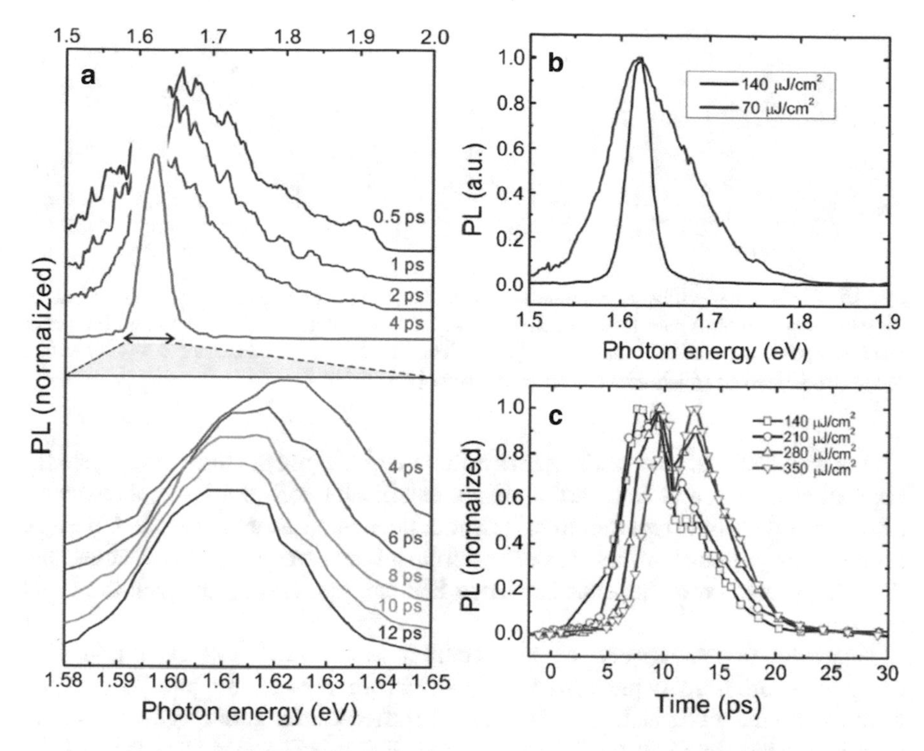

Fig. 3.5 Effects of carrier thermalisation in free-carrier optical gain in $CH_3NH_3PbCl_xI_{3-x}$ PTFs [8]. **a** OKG spectra at different instants, showing that the onset of free-carrier ASE is hot-carrier thermalisation limited. **b** The onset of narrowband ASE occurring at the same peak position, suggesting avalanche band-edge recombination mechanisms. **c** Partial recovery of ASE after the first ASE event caused by newly thermalised hot carriers, as seen by the later time "shoulder"

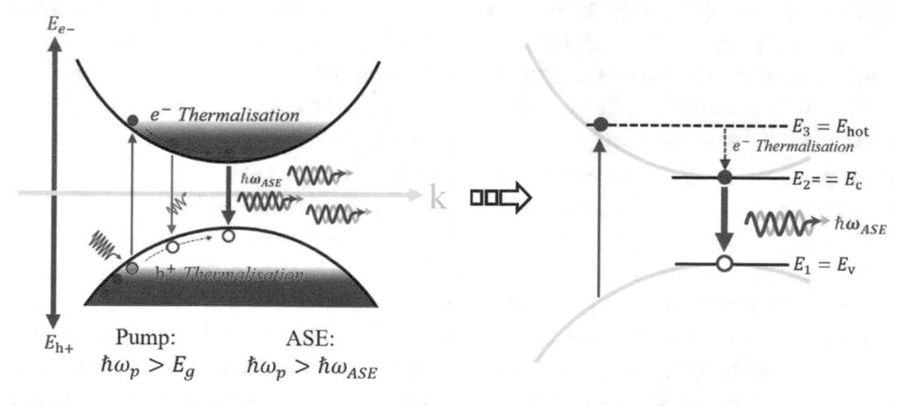

Fig. 3.6 An illustration of the free-carrier optical gain mechanism. Population inversion of electron and hole carriers occur along their band-edges, which makes it fundamentally thermalisation-limited. Hence, the free-carrier gain mimics a three-level system scheme

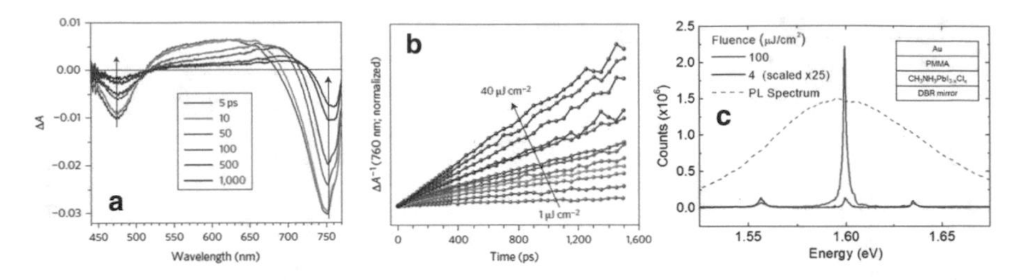

Fig. 3.7 Bimolecular radiative recombination indicative of photogenerated free-carriers in $CH_3NH_3PbI_3$ thin films shown in **a** photobleaching (PB) dynamics in TA spectra and **b** a linear relationship between $(\Delta A)^{-1}$ and time t [12]. **c** The first free-carrier Perovskite VCSEL using spincoated $CH_3NH_3PbCl_xI_{3\text{-}x}$ PTF as its active layer [11]

The drawbacks of free-carrier gain are that in mimicking the three-level systems, population inversion of free-carriers is not as efficient, where a large photogenerated carrier density is required to bring about population inversion at the CB edge. Secondly, any premature hot carrier recombination serves as a loss channel that (Fig. 3.6, green arrow) impedes bandedge buildup, thus raising the gain threshold further.

Generally, free-carrier optical gain occurs in systems with low exciton binding energies, whose primary photogenerated carriers are free-electron and holes. Apart from the common $CH_3NH_3PbI_3$ PTFs [10–12], free-carrier gain mechanisms have been suggested for $CsPbBr_3$ PNWs [13, 14], PNCs [15] and microspheres [16]. Formal verifications of free-carrier recombination at below threshold excitations in $CH_3NH_3PbI_3$ PTFs have been studied by Manser and Kamat et al. [12] and Ghanassi et al. [17] via pump-probe experiments. As shown in Fig. 3.7a, b, photobleaching (PB) dynamics is ascribed as second-order bimolecular kinetics [12] and a strong **linear** relationship between the reciprocal differential absorbance $(\Delta A)^{-1}$ with time [17] are indicative of free carrier state-filling [12]. These two mutual verifications can be understood if we consider a system with dominant bimolecular recombination in the low fluence regime. Its rate equation is modelled as $\frac{dn}{dt} = -Bn^2$, which gives a carrier density solution of $n(t) = \frac{n_0}{1+n_0Bt}$. Since the probe's change in absorbance ΔA is proportional to PB its carrier density n, we see that $(\Delta A)^{-1} \propto n^{-1} \propto Bt$. Here, the gradient B is the corresponding bimolecular radiative recombination rate. In the high fluence regime, where strong non-radiative auger recombination (trimolecular) is expected to dominate, Manser and Kamat et al. observed that bimolecular radiative recombination of free carriers still dominated even above 40 μJcm^{-2} [12]. From a temporal point of view, this meant that free-carrier ASE could still occur because the sub-ps bandedge optical buildup, hot carrier thermalisation lifetimes outcompete the relatively slower auger recombination with ~10^1 ps timescales. The first free-carrier Perovskite VCSEL was constructed by Deschler et al. in 2014, as shown in Fig. 3.7c, where the end-reflector is a DBR coupled to a gold (Au) output coupler [11]. PMMA was used as a protective layer to avoid physical damage to the Perovskite active layer during device fabrication [11].

Another approach to obtain free-carrier optical gain is to bring the carrier density in excitonic Perovskite systems beyond its Mott-transition density, where photogenerated excitons actively dissociate back to free-carriers in the high pumping density regime. For instance, while many argued optical gain mechanisms in $CsPbBr_3$ PNCs to be trionic or biexcitonic, Geiregat et al. was the first to instead, propose free-carrier ASE mechanisms in weakly confined $CsPbBr_3$ PNCs with edge-lengths ~12.7 nm [15]. In this work, it was reported that despite room temperature excitonic absorption by their ensembles, the ASE is found to be thermalisation limited and occurs at threshold carrier densities situated above the Mott-transition density limit. To understand the role of hot-carrier cooling in $CsPbBr_3$ PNC films, Geiregat et al. compared differences in carrier dynamics between ensembles pumped using 400 and 500 nm femtosecond pulses. Mainly, the threshold carrier density of ASE after accounting for difference in absorption cross-sections remained around the same for both cases ($n_{th} \sim 1.6 - 1.8 \times 10^{18} \text{cm}^{-3}$), yet a higher intrinsic gain coefficient g_i and almost instantaneous onset of ASE were observed in the resonantly pumped case using 500 nm pulses. On the other hand, the 400 nm pumped case exhibited a much lower value of g_i with a delayed ASE onset. Such observations strongly indicate that the ASE occurring in their weakly confined $CsPbBr_3$ PNCs are thermalisation-limited, which point towards a free-carrier based mechanism [15]. Although not explained in the report, g_i measured in the 500 nm resonantly pumped case was higher than at 400 nm pump case possibly because of the premature hot carrier recombination loss that impedes the accumulation of population inversion at the bandedge that lowers the effective gain in the ensemble.

To further support his initial claims, Geiregat et al. proposed two other proofs, by comparing the acquired n_{th} relative to (I) the ensemble's size-distribution and (II) mott-transition carrier density [15]. In this report, the observed ASE is shown to possess an unchanging $n_{\text{th}} \sim 1.6 \times 10^{18} \text{cm}^{-3}$ with crystallite size that is similar to its bulk-like counterparts that is given by [15]:

$$n_{\text{th(FC)}} \approx \frac{3}{2}\left[\frac{M}{M_r}\right]^{3/4}\left[\frac{2\pi M_r k_B T}{h^2}\right]^{3/2} \tag{3.6}$$

as opposed to a biexcitonic gain mechanism that exhibit volume-scaling of $n_{\text{th(XX)}} = \frac{\langle N \rangle}{V_{\text{NC}}}$ [18]. Here, M and M_r denote the total and reduced exciton masses and T is the carrier temperature. For the second proof, Geiregat and co-workers produced a numerical calculation of the mott-transition and free-carrier optical gain threshold carrier densities based on the density functional theory calculations using the generalised gradient approximation (DFT-GGA) approach, with and without considering the van der Waals interaction (VDW), as shown in Fig. 3.8c [15]. Essentially, the Mott-transition characterises the threshold carrier density where excitons would actively dissociate back to its free-carriers as a result of carrier-carrier Coulomb screening effects [18].

The Mott-transition density of a system is as a function of excitonic Bohr radius r_B and binding energies Δ_X that is given by [18]:

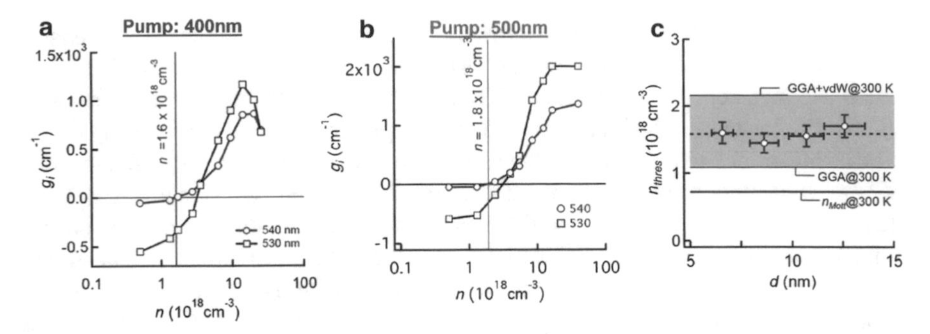

Fig. 3.8 Characterisation of free-carrier optical gain mechanisms in weakly confined $CsPbBr_3$ PNCs [15]. A comparison of intrinsic gain coefficient g_i between **a** 400 nm and **b** 500 nm (resonant) pumping conditions revealed that the former suffered from lower value of g_i and a delayed onset of ASE, suggesting the critical role of hot carrier thermalisation effects. **c** A numerical calculation based on Density Functional Theory in the Generalised Gradient Approximation approach (DFT-GGA), showing the Mott-transition carrier density (blue line) and the expected threshold carrier density for free-carrier ASE (with and without considering van der Waals interactions) in their $CsPbBr_3$ PNCs

$$n_{\text{mott}} \approx 0.028\left[\frac{k_B T}{r_B \Delta_X}\right] \tag{3.7}$$

By using Eq. (3.7), a Mott-transition density of $n_{\text{mott}} \sim 4.7 \times 10^{17}\text{cm}^{-3}$ for $CsPbBr_3$, with $r_B = 3.5\,\text{nm}$ and $\Delta_X = 36\,\text{meV}$ can be determined, where the acquired $n_{th} > n_{\text{mott}}$. Thus, this indicates that light amplification in $CsPbBr_3$ PNCs is a result of free-carrier accumulation at its CB edge, where an initially dense exciton carrier-carrier screening resulting in active exciton dissociation and leading to thermalisation-limited free-carrier gain mechanisms [15].

Finally, it is important for readers to notice that early reports studying optical properties of $CH_3NH_3PbI_3$ PTFs may assign the radiative recombination to excitons [19] instead of free carriers [12]. This discrepancy could be due to the morphological quality of PTFs. In the early days of the field, $CH_3NH_3PbI_3$ PTFs were morphologically imperfect, with ~10^2 nm grain sizes and pinholes aplenty with insufficient film coverage on substrates. As such, these films possessed optical properties resembling larger PNCs and typically possessed relatively higher room-temperature excitonic binding energies than its bulk-like large-grain film counterparts. Thus, it is possible for the primary excitations in $CH_3NH_3PbI_3$ PTFs prepared in the early days, to be more excitonic under room-temperature conditions.

3.1.3 Biexciton Optical Gain

Biexcitonic optical gain occurs typically in excitonic systems, where the population inversion of biexcitons and biexcitonic stimulated emission occurs, as illustrated in Fig. 3.9. Biexcitons are formed under the influence of attractive exciton-exciton interactions under exciton-dense conditions. As such, the biexciton's total energy E_{XX} is less than 2 individual excitons $2E_X$ by the binding energy Δ_{XX} given by [9]:

$$\Delta_{XX} = 2E_X - E_{XX} \tag{3.8}$$

Evidently, the biexciton gain is also analogous to a "quasi" three-level system gain mechanism, where the pump generated "quasi-level" $2E_X$ forms the short-lived top-most level, that is succeeded by ultrafast exciton-exciton attractive binding to the metastable biexcitonic state E_{XX}. Upon fulfilling population inversion of biexcitons relative to excitons, biexciton gain occurs, where avalanche stimulated emission of biexcitonic recombination produces photons with energy $\hbar\omega_{XX}$. In the process, this leaves behind an exciton with energy E_X. In using Eq. (3.8) and the fact that $\hbar\omega_{XX} = 2E_X - \Delta_{XX}$, we can express Δ_{XX} in terms of experimental observables E_X and $\hbar\omega_{XX}$ as follows:

$$\Delta_{XX} = E_X - \hbar\omega_{XX} \tag{3.9}$$

Equation (3.9) tells us that the biexciton binding energy can be deduced from the red-shift of biexciton PL peak relative to its exciton PL peak via PL spectral deconvolution. In practice, materials possessing large Δ_{XX} are ideal for realising low threshold biexcitonic gain because it decreases the extent of spectral overlap between biexciton and exciton PL bands and that biexciton ASE can occur with minimal re-absorption losses [20]. In this case, Eq. (3.9) can be used to calculate Δ_{XX} by directly taking the exciton and biexciton ASE peak difference. Conversely,

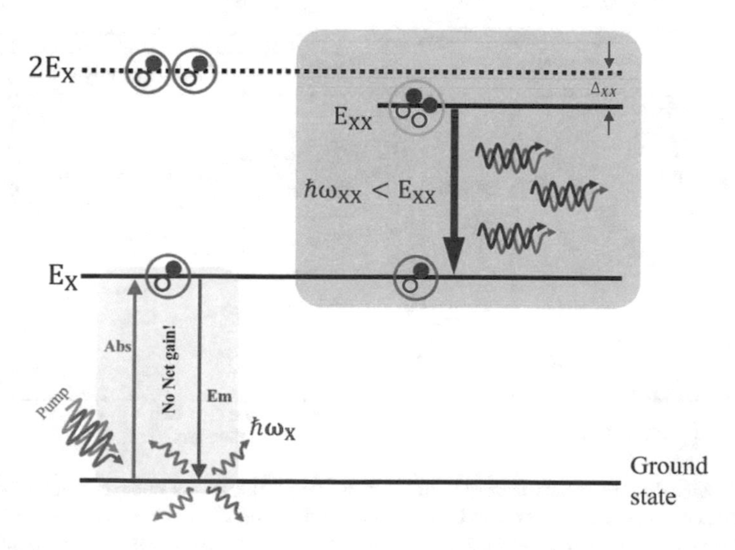

Fig. 3.9 Schematic overview single exciton (gray box) and biexciton gain (orange box) mechanisms in the excitonic framework

for systems with small Δ_{XX}, material re-absorption is still strongly competing with biexcitonic buildup, which arbitrarily raises the gain threshold where the biexciton ASE further red-shifts [20]. In this case, both PL bands are strongly overlapping and the true value of Δ_{XX} must be acquired through spectral deconvolved PL bands instead of direct subtraction with the ASE band. Large Δ_X ~50 meV [2] to ~80 meV [9] was reported in $CsPbBr_3$ PNCs, which is sufficiently displaced out of its excitonic linewidth (~100 meV). With reference to Fig. 3.9 again, the reason that natural single exciton gain cannot happen in excitonic systems is because of the exact compensation between absorption and stimulated emission rates (recall from Sect. 1.2.1) in such a two-level system.

Similarly, proof of biexciton carriers in the system is crucial for validating biexciton gain mechanisms. In 2015, Wang et al. formally characterised the existence of biexcitonic PL in $CsPbBr_3$ PNCs excited via ~ns pulses under 10 K conditions, as shown in Fig. 3.10 [2]. Here, the rationale of using ~ns pulses instead of ~ fs pulses is to virtually increase the ASE threshold to ensure that the characterisation would not include ASE in the spectra. In addition, ultralow temperatures at 10 K was employed

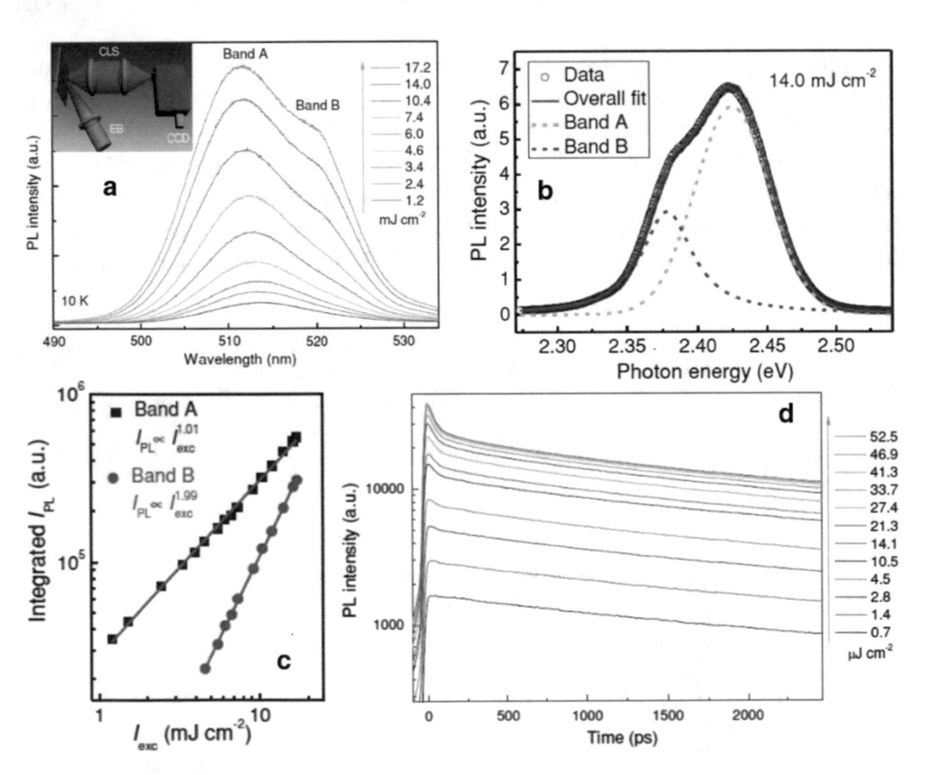

Fig. 3.10 Biexciton PL characterization in $CsPbBr_3$ PNCs under ~ns pulsed excitations at 10 K [2]. **a** Fluence dependent PL spectra. **b** Spectral deconvolution of bands A and B via a two-peak pseudo-Voigt fitting and **c** the fluence dependent scaling of deconvolved peaks. **d** Fluence dependent PL kinetics under ~fs pumping

to minimize spectral broadening from phonon interactions for better spectral resolution [2]. Figure 3.10a shows that fluence dependent PL spectra where a second red-shifted peak develops. Figure 3.10b shows the spectral deconvolution using a two-peak pseudo-Voigt function fitting. Figure 3.10c shows the fluence dependent trends of spectral deconvolution, where the red-shifted band B manifesting in the higher fluence regimes were found to scale quadratically with fluence and correlates to biexciton recombination mechanism. The quadratic scaling is because the biexciton radiative recombination rate is proportional to the probability of either exciton recombining (i.e.$P(\hbar\omega_{XX}) \propto P(X) \times P(X)$). On the other hand, the proof of biexciton signatures is not as straightforward in TR-PL kinetics as in steady-state PL spectra. With increasing fluence, the exciton-dense media is not only susceptible to biexciton formation from attractive binding but also susceptible to exciton-exciton annihilation (XXA). XXA is an ultrafast bimolecular [21] non-radiative auger recombination process in which an exciton recombines non-radiatively and transfers this energy to the other exciton [22]. Thus, as shown in Fig. 3.10d, the early-time fast component with tens of ps lifetime is an undeniable proof of XXA and its contribution increases in the kinetics with increasing fluence. XXA is undesirable, as it can compete with the formation of biexciton carriers necessary to populate $|XX\rangle$. At above threshold excitation conditions, a typical TR-PL spectrograph would resemble that shown previously in Fig. 3.2a, with time-resolved PL and ASE kinetics shown in Fig. 3.2b [1]. In the case of $CH_3NH_3PbBr_3$ PNCs (similar to $CsPbBr_3$ PNCs), the biexcitonic ASE kinetics is extremely short-lived in comparison to its PL decay trace. Furthermore, its ASE kinetics is dominated by an initial fast component, which is assigned as biexciton gain decay lifetime τ_g and the later component assigned as τ_{MPE} that includes all combinations of exciton-biexciton auger decays. Here, the PL kinetics (blue) of $CH_3NH_3PbBr_3$ PNCs are similar to the PL kinetics of $CsPbBr_3$ PNCs shown in Fig. 3.10d, where the fast component at ~34 ps is assigned as the XXA process [1]. Figure 3.2d illustrates the complete picture detailing the biexciton gain build-up and decay in both time and fluence regimes.

At below threshold pumping conditions, XXA occurs without substantial biexciton build-up. Once above threshold pumping conditions, a delayed onset of biexciton build-up $\tau_{\text{P.Inv}}$ occurs because it precedes the XXA losses τ_{X-X} [1, 2]. In the emission process, the gain lifetime is shortened by general MPEs involving all combination of exciton and biexciton auger processes that rapidly reduces the biexciton optical build-ups. For these reasons, biexciton gain are auger-limited gain mechanisms [2, 20], where and MPE shortens the biexciton gain lifetime τ_g, as illustrated over a timeline and fluence-line in Fig. 3.2d. Fast XXA rates has been measured in $CsPbBr_3$ PNCs at $\gamma = \left(7 \times 10^{-7}\right) \text{cm}^3\text{s}^{-1}$ [21]. XXA characterisation in biexciton gain studies can be found also in organic materials [23], transition metal dichalcogenides [24] and chalcogenide QDs [5].

Biexciton gain mechanisms are well-studied in chalcogenide QDs by Klimov et al. and co-workers, who derived the biexciton gain threshold in the electronic framework. Generally, the number of excitons N photogenerated in an ensemble during excitation follows a Poisson distribution, which is defined as:

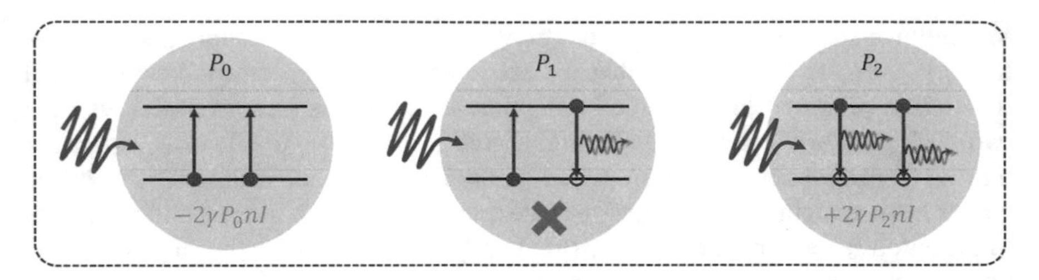

Fig. 3.11 Biexciton gain mechanism of neutral PNCs in the electronic framework. Here, biexcitons are visualised as two pairs of electrons and holes. This illustration is applied in pump-probe studies, where probe photons can either be absorbed or induce stimulated emission transitions

$$P_N = \frac{e^{-\langle N \rangle} \langle N \rangle^N}{N!} \tag{3.10}$$

where P_N is the probability of finding a QD that contains N excitons and $\langle N \rangle$ denotes the ensemble averaged number of excitons per QD. In practice, $\langle N \rangle$ is a useful quantity that is defined as:

$$\langle N \rangle = \frac{\sigma_{abs} F}{\hbar\omega} \tag{3.11}$$

where σ_{abs} is the linear absorption cross-section, whose value is affected by pump energy $\hbar\omega$. From Eq. (3.11), since $\langle N \rangle \propto F$, it provides as a representation of pump-efficiency after accounting for a material's absorption cross-section. According to Eq. (3.10), the statistics tell us that if $\langle N \rangle = 1$, the probabilities of finding unexcited P_0, singly excited P_1 and multiexcited $P_{N>2}$ are non-zero, as shown in Fig. 3.11. Now, by further defining γ and n to be the probability of excitonic recombination and the number density of neutral PNCs, each type of QD with N excitons can change the seed light intensity I as follows:

$$\Delta I = -2\gamma P_0 nI + 0 + 2\gamma P_2 nI + 2\gamma P_3 nI + \cdots = 2\gamma nI \left[-P_0 + \sum_{N=2}^{\infty} P_N \right] \tag{3.12}$$

The first term $-2\gamma P_0 nI$ in Eq. (3.12) corresponds to light intensity loss due to absorption by unexcited QDs, where the factor of 2 accounts for the possibility of two simultaneous absorption events by VB edge electrons. The second term of zero comes from the transparency due to singly-excited QDs, due to the exact compensation between stimulated emission and absorption. The third term onwards correspond to contributions of gain from multi-excited QDs, where $N \geq 2$. Similarly, the factor of 2 for these higher ordered terms accounts for the possibility of inducing two band-edge stimulated emission contributions that is strictly limited by Pauli's exclusion principle. From here, we can see that since $\langle N \rangle \propto F$, the biexciton gain occurs at a relatively high threshold in an exciton dense media. The biexciton gain threshold can then be derived by letting $\Delta I = 0$, which gives us:

$$\left[-P_0 + \sum_{n=2}^{\infty} P_n\right] = 0 \tag{3.13}$$

In using Eq. (3.10) with (3.13), the biexciton gain threshold in terms of $\langle N \rangle$ is therefore calculated to be:

$$\langle N \rangle_{\text{th(XX)}} = 1.15 \tag{3.14}$$

Our result in Eq. (3.14) is not surprising, as we intuitively expect the ensemble average to be more than 1 exciton, where $\langle N \rangle_{\text{th(XX)}} > 1$. In practice, this formalism applies to pump-probe experiments, where Eq. (3.12) tabulates all terms comprising of probe absorption and probe-induced stimulated emission.

Very recently, two-dimensional electron spectroscopy (2DES) studies unveil that the gain mechanism in $CsPbBr_3$ PNCs can be ascribed to biexcitons [9]. Importantly, the results showed that biexcitonic formation signatures can be detected either by (I) excited state absorption (ESA) in singly excited PNCs or (II) directly via exciton-exciton interactions in doubly excited PNCs accompanied by a delayed ASE onset [9]. Figure 3.12a shows the fluence dependent steady-state PL spectra, where a red-shifted (~76 meV) ASE band is observed. Figure 3.12b shows the 2DES absorptive spectra at $t = 5$ ps under 25 μJcm^{-2} pumping conditions [9]. By resolving the 2DES absorptive kinetics at 2.39 eV (pink square) as shown in Fig. 3.12c, fluence dependent probe signals showed a gradual crossover from negative signal (ESA) to positive signal (stimulated emission) with increasing pump fluence [9]. Wei et al. explained that the short-lived ESA originates from the excitonic transitions up to filling the biexcitonic states (i.e. $|X\rangle \rightarrow |XX\rangle$), as shown in Fig. 3.12d.

Importantly, the crossover meant that the onset of ASE in $CsPbBr_3$ is delayed due to the short-lived initial ESA associated with biexciton formation resulting from attractive carrier interactions [9]. On the other hand, single exciton gain is ruled out as it would instead, have shown simultaneous ESA and SE signals. For singly excited ($N = 1$) PNC, the ESA_X effectively competes with SE_X while doubly excited ($N = 2$) PNCs may directly covert into biexcitons and contribute to biexcitonic ASE (SE_{XX}). Figure 3.12e shows fluence dependent probe signals at 2.39 eV (pink squares) and 2.53 eV (green circles). The positive and superlinear relation of 2.39 eV probe signal with fluence suggest a multiexcitonic interaction regime that supports biexcitonic gain mechanisms [9]. Biexcitonic ASE onset delays are expected to shorten with fluence as more PNCs are doubly-excited, thereby inhibiting the initial ESA process for biexcitonic formation in singly excited PNCs [9]. By global analysis of the entire 2DES data, Wei et al. determined that auger recombination with fitted $\tau_{au} \sim 80$ ps is much slower and outcompeted by the ultrafast ESA with lifetimes <2 ps, which justifies the observation of net gain in their $CsPbBr_3$ PNCs. Interestingly, Wei et al. noticed that the probe energy difference between exciton photobleaching

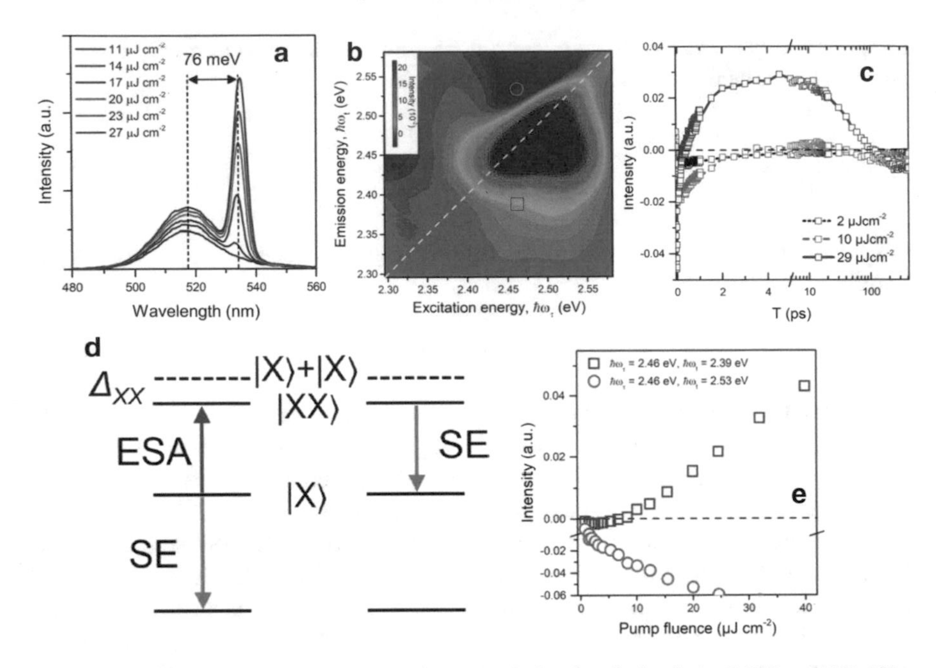

Fig. 3.12 2DES absorptive spectral and kinetic analysis of optical gain in $CsPbBr_3$ PNCs [9]. **a** Fluence dependent steady-state PL spectra. **b** 2DES absorptive spectrograph at t = 5 ps excited at above ASE threshold of 25 μJcm^{-2}. **c** Fluence dependent 2DES absorptive kinetics resolved at 2.39 eV. **d** Schematic illustration biexciton formation via ESA in singly excited PNCs, that is competed by excitonic SE_X. SE_{XX} is the biexcitonic ASE transition. **e** Fluence dependent 2DES probe signals at 2.39 eV (pink squares) and 2.53 eV (green circles)

(PB_X) and SE_{XX} signatures matches well with the 76 meV ASE red-shift shown in Fig. 3.12a. This suggests that the ASE red-shift is a direct result of attractive Δ_{XX} that creates biexciton states that are sufficiently spectrally displaced from the linear absorption transition of excitons by unexcited PNCs. In order to support this claim, Wei et al. studied the effects of tuning the ASE spectral redshift relative to the excitonic PL's bandwidth Γ. Next, f is defined as the effective ratio of PNCs participating in linear absorption for a transition displaced by Δ_{XX} relative to E_X, as shown in Fig. 3.13a. By modifying Eq. (3.12), the ratio f can be appended into the formalism to give:

$$-\Delta\alpha = \alpha_0 f - \alpha = 2\gamma f - \left[2\gamma f P_0 + \frac{1}{2}\gamma P_1 - \frac{1}{2}\gamma f P_1 - 2\gamma P_2\right] \quad (3.15)$$

where $\Delta I \propto -\alpha$ is the detected probe signal measuring optical gain and γ is the absorption and emission rates. The terms in the parenthesis correspond to (I) the detuned absorption by unexcited PNCs, (II) the ESA of singly excited PNCs in biexciton formation, (III) SE_X of detuned excitons and (IV) the direct SE_{XX} contribution from doubly-excited PNCs, respectively. Equation (3.15) aims to account for

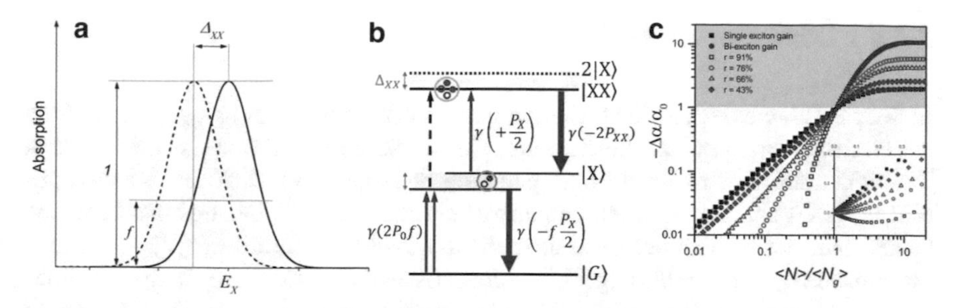

Fig. 3.13 Role of attractive carrier potentials and spectral overlap in biexcitonic gain mechanism of $CsPbBr_3$ PNCs [9]. **a** Schematic illustrating biexciton transitions displaced by Δ_{XX} relative to linear excitonic transition E_X contains spectral overlap with a ratio f, of PNCs susceptible to reabsorption. **b** Modified schematic accounting for spectral-overlap induced transitions that is previously ignored in the formalism. **c** Projected probe signals as a function of normalised pump density $\frac{\langle N\rangle}{\langle N\rangle_g}$ for different biexcitonic transition detuning term r due to attractive binding. Gray region indicates ASE threshold. *Note* "Bi-exciton" term used in this report describes two non-interacting excitons and differs from our description of biexcitons bound system

transition "blurring" due to the close proximity in transition energies observed in pump-probe experiments. Note that the factors of $\frac{P_1}{2}$ and $f\frac{P_1}{2}$ in terms (II) and (III) correspond to the corrected probabilities that the single exciton will either undergo ESA or SE_X. Furthermore, since terms (II) and (IV) are probe-induced resonant processes, they do not include the correction term f. The term $\alpha_0 f = 2\gamma f$ describes the probe's detection for "detuned" linear absorption in the absence of pump, where $\alpha_0 \propto 2\gamma$. All these processes are illustrated in Fig. 3.13b. By further defining the extent of spectral overlap r as the ratio between Δ_{XX} and Γ:

$$r = \frac{\Delta_{XX}}{\Gamma} = \sqrt{-\frac{\ln f}{4\ln 2}} \tag{3.16}$$

the effect of tuning the values of r for fixed exciton PL width Γ relative to normalised pump $\frac{\langle N\rangle}{\langle N\rangle_g}$ is shown in Fig. 3.13c. Here, the extreme at $r \rightarrow 0$ corresponds to non-interacting carriers while $r \rightarrow 1$ corresponds to a complete spectral shift such that $f \rightarrow 0$ with negligible overlap. Importantly, while excitonic and **non-interacting** two-excitonic (termed bi-excitonic in this report) gain behaviours are found to be rather similar, tuning the overlapping ratio $0.43 < r < 0.91$ showed that as r increases, deviation from the single exciton gain becomes more pronounced, with the probe signal growing more steeply that indicates greater net modal gain. In their $CsPbBr_3$ PNCs, a value of $r = 0.88$ was experimentally obtained, which estimates an attractive $\Delta_{XX} \sim 88$ meV that agrees well to the estimated value of 76 meV in Fig. 3.12a.

3.1.4 Trion Optical Gain

In Sect. 3.1.3, we derived the biexciton gain threshold to be $\langle N \rangle_{\mathrm{th(XX)}} \sim 1.15$ from the electronic framework. Furthermore, we discussed that the major drawback of biexciton gain is its relatively high pumping threshold $\langle N \rangle \propto F$ and its susceptibility to XXA and higher-ordered multiparticle auger effects. On the other hand, the trionic gain mechanism occurs at theoretically lower thresholds of $\langle N \rangle < 1$ and is less affected by undesirable auger limitations. Essentially, trions are charged excitons that was first predicted in 1958 [25] and observed much later in $CdTe/Cd_{1\text{-}x}Zn_xTe$ quantum wells in 1993 [26] and subsequently in other monolayer transition metal dichalcogenides [27], nanotubes [28, 29], nanoplatelets [30] and nanocrystals [31]. Early reports suggest that trions are typically formed when an exciton binds with an additional "stray" charge [32] that results either from surface-defect induced photocharging [33, 34], multiexcitonic auger recombination [35] and/or auger-ionization [36]. In the **low fluence limit** $\langle N \rangle \ll 1$, positive trions X^+ are formed in negatively photocharged PNCs due to unpassivated surface-dangling Pb-ions acting as strong electron traps [37, 38]. In the **high fluence limit** $\langle N \rangle > 1$, trions can also intrinsically form in neutral PNCs as a result of auger-ionization, where strong auger recombination rates rapidly dissociate and eject charges with high excessive energies [38, 39]. Post-synthetic passivation treatments with $PbBr_2$ [6] or NaSCN (sodium thiocyanate) [38] have been shown to almost completely suppress the formation of electronic traps while stirring the colloidal PNC during photoexcitation is shown to suppress auger-induced trion formation [40]. Trion gain mechanism is an interesting topic because trions were originally thought to be undesirable carriers. Preliminary isolated-dot studies held trions responsible for PL intermittency (photoblinking) [41, 42] in light emitting materials and are expected to lower efficiency in LEDs and laser devices. Yet later, scientists are able to capitalise on the net half-spin properties of trions, leading to developments in ultrafast photocurrent switches [43], electronic spin manipulation [43], quantum computing [44], LEDs [45] and most importantly, lasers [6, 20, 46].

By modifying the electronic framework and using Poisson statistics describing the PNC ensemble's excitation distribution, the threshold conditions for trion gain in PNCs can be derived analogously [6]. Regardless of photocharging mechanism, photoexcitation of an ensemble of singly-charged PNCs results in unexcited (P_0), singly excited (P_1), doubly excited (P_2) PNCs and etc. ($p_{n>2}$), as shown in the schematic of Fig. 3.14a.

Next, we define γ and n_c as the recombination probability and the number density of negatively charged PNCs in the ensemble (per volume). In assuming trionic spontaneous emission with seeding intensity I is present in the ensemble media, the total change in light intensity as a result of unexcited and excited PNCs are given by [6]:

$$\Delta I = -\gamma P_0 n I + \gamma P_1 n I + 2\gamma P_2 n I + 2\gamma P_3 n I + \dots$$

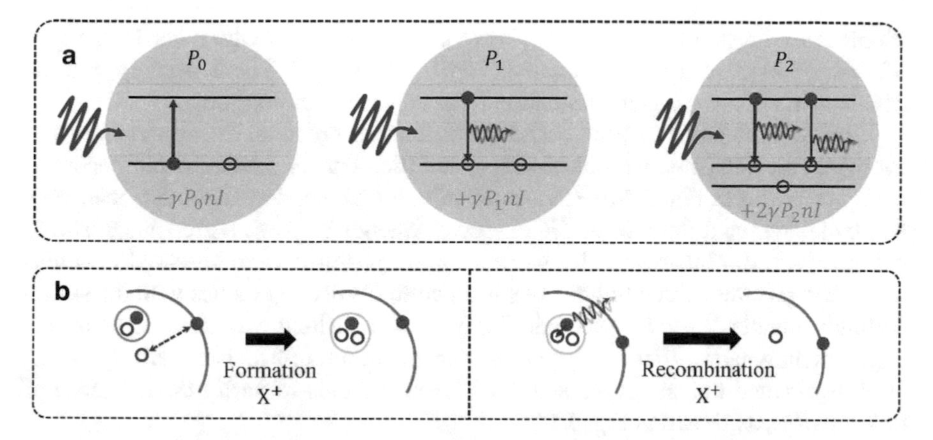

Fig. 3.14 **a** Illustration of positive trion X^+ gain mechanism in the electronic framework. **b** Schematic of X^+ carrier formation and recombination. The converse is true for negative trions in positively charged PNCs

$$= \gamma n I \left[-P_0 + P_1 + 2 \sum_{n=2}^{\infty} P_n \right] \tag{3.17}$$

In contrast to Eq. (3.12) for biexciton gain, singly excited PNCs do contribute to trion gain as population inversion is naturally satisfied, with contribution term $\gamma P_1 n I$. However, multi-excited PNCs generally contribute with similar term $2\gamma P_n n I$ as a consequence of stimulated emission bottleneck imposed by the Pauli's exclusion principle. Since net gain occurs when $\Delta I > 0$, the threshold condition is therefore given by $\Delta I = 0$:

$$\left[-P_0 + P_1 + 2 \sum_{n=2}^{\infty} P_n \right] = 0 \tag{3.18}$$

With reference to Eqs. (3.10), (3.18) is solved analytically, which gives the threshold for trion gain to be [6]:

$$\langle N \rangle_{\text{th}(X^+)} = 0.58 \tag{3.19}$$

Comparing our results between biexciton and trion gain thresholds in Eqs. (3.14) and (3.19), respectively, it is clear that the accuracy of the result requires the inclusion of all $P_{n>2}$ contributions, despite them occupying diminishing portions in the Poisson distribution of the ensemble. In addition, the significant reduction in pumping threshold from $\langle N \rangle_{\text{th}} = 1.15 \rightarrow 0.583$ in trionic gain originates from the active contribution of singly excited PNCs P_1. In practice, this formalism can be applied in pump-dependent pump-probe experiments, with the probe photon resonant to the trionic transition acting as the input seed. From Fig. 3.4, Wang et al. acquired a trionic ASE threshold of $\langle N \rangle_{th} = 0.62 (\text{or} F_{th} = 1.2\,\mu\text{Jcm}^{-2})$ with much longer gain lifetime ~660 ps in the $PbBr_2$-treated $CsPbBr_3$ PNCs [6]. These strongly indicate trionic gain with lesser gain dissipation from auger-recombination.

Similarly, the proof of trion carriers' existence is essential to support claims of trion gain mechanisms. An illustration of the trion carrier formation and recombination is shown in Fig. 3.14b. Conventionally, trionic carriers have also been verified by taking the difference of TR-PL decay kinetics between stirred and unstirred colloids [47]. In Makarov et al.'s work, trionic signatures were observed with lifetime ~235 ps after subtracting the unstirred colloid's decay kinetics with the stirred colloid's monoexponential excitonic decay kinetics, which was assigned as trionic auger recombination lifetime instead of carrier lifetime [40]. Alternatively, Wang et al. performed fluence dependent TR-PL experiments to verify the existence of trion carriers, as shown in Fig. 3.15.

In this case, proper subtraction can only be done after ensuring that the long-time PL kinetics tail corresponding to excitonic decay are mutually normalised across all fluence-varied kinetics [6]. As shown in Fig. 3.15a, subtracting the PL decay kinetics between $\langle N \rangle = 0.052$ and $\langle N \rangle = 0.006$ yields the blue decay trace, which is fitted and assigned as trionic radiative lifetime at $\tau_{X^{+}(r)} = 1.63\,\text{ns}$. In Fig. 3.15b, it can also be seen that with further increase in fluence till $\langle N \rangle = 0.42$, the subtracted kinetics in blue shows an even faster component with lifetime $\tau_{XX(aug)} = 320\,\text{ps}$ and was assigned as biexcitonic auger recombination lifetime.

Seeing that the assignment of subtracted decay components are often debated between trionic and biexcitonic carriers (which impacts the assignment of trionic gain), one way to distinguish them is to consider the so-called statistical scaling of decay rates by excitons, trions and biexcitons. For a system with quantum yield Φ_f, the TR-PL measured lifetime τ_X relates to its true radiative lifetime via the relation $\tau_{r(X)} = \frac{\tau_X}{\Phi_f} = \frac{1}{k_{r(X)}}$. Since the excitonic radiative decay rate depends on the electron and hole carrier densities such that$k_{r(X)} \propto n_e n_h$, then we can generalise the decay rate for an Mth order exciton:

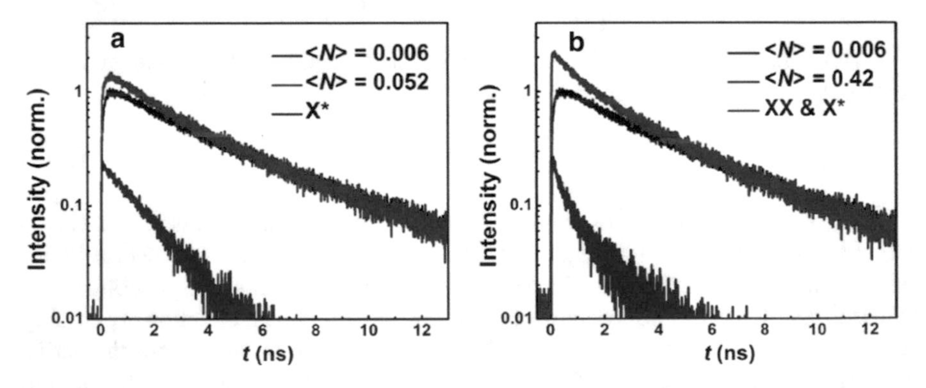

Fig. 3.15 Fluence dependent TR-PL kinetics of $PbBr_2$-treated $CsPbBr_3$ PNC films normalised at $t = 8\,\text{ns}$ in the **a** very low and low fluence regime and **b** very low and high fluence regime. The blue decay trace is the result of subtraction between the respective two decay kinetics in each figure

$$k_{r(MX)} \propto (Mn_e)(Mn_h) \propto M^2 k_{r(X)} \tag{3.20}$$

Since $k_{r(MX)} = \frac{1}{\tau_{r(MX)}}$, we deduce from Eq. (3.21) that the Mth order excitonic radiative lifetime scales with its excitonic radiative lifetime as follows:

$$\tau_{r(MX)} = \frac{\tau_{r(X)}}{M^2} \tag{3.21}$$

where $\frac{1}{M^2}$ is known as the order scaling factor. For trions, the radiative rate is$k_{r(X^+)} \propto (n_e)(2n_h) \propto 2k_{r(X)}$, which yields a scaling of $\tau_{r(X^*)} = \frac{\tau_{r(X)}}{2}$. By extension of our result in Eq. (3.22), biexcitons would have a lifetime of $\tau_{r(XX)} = \frac{\tau_{r(X)}}{4}$. Therefore, knowledge of the material's $\tau_{r(X)}$ and Φ_f is essential to provide the appropriate assignments of the observed components. For most Perovskite systems, where $\tau_{r(X)}$ ~ ns, we expect $\tau_{r(X^*)}$ in PNCs to also occur in the ~ns timescales.

Finally, it should be remarked that the extent of trion localisation and the energetics behind its gain mechanisms are still poorly understood. Recently, it was reported that selective generation of bound trions in Single-Walled Carbon NanoTubes (SWCNTs) showed significantly higher trionic binding energies and emission intensities than its intrinsic free trions [48]. Secondly, the origins of optical gain in $CsPbBr_3$ PNCs are widely debated and assigned to be due to trions [6], free carriers [15] and biexcitons [9]. However, all these reports could be valid, and conclusions could differ due to differences in material quality such as synthetic and storage temperature and conditions.

3.1.5 Single Exciton Optical Gain

Earlier in Fig. 3.13, we discussed that singly-excited PNCs cannot produce net optical gain due to equal compensation between absorption and stimulated emission rates. In chalcogenide QDs, core–shell structures such as CdS/ZnSe were morphologically engineered in order to capitalise on the pump-induced transient stark effect that could temporarily unbalance the stimulated emission and absorption rates [5]. Specifically, since CdS/ZnSe follows a type II core–shell energetics, excitons generated are spatially charge-localised which produces a transient electric field directed towards the CdS core. As shown in Fig. 3.16, this further results in a transient stark effect that lifts the electronic degeneracies, such that the absorption line is sufficiently detuned from the stimulated emission line [5]. The single exciton gain threshold in terms of $\langle N \rangle$ can be derived analogously, except that there will be **no contributions from multiexcitons**. With reference to Fig. 3.16, the net change in light intensity is given by:

$$\Delta I = -2\gamma P_0 n I + \gamma P_1 n I \tag{3.22}$$

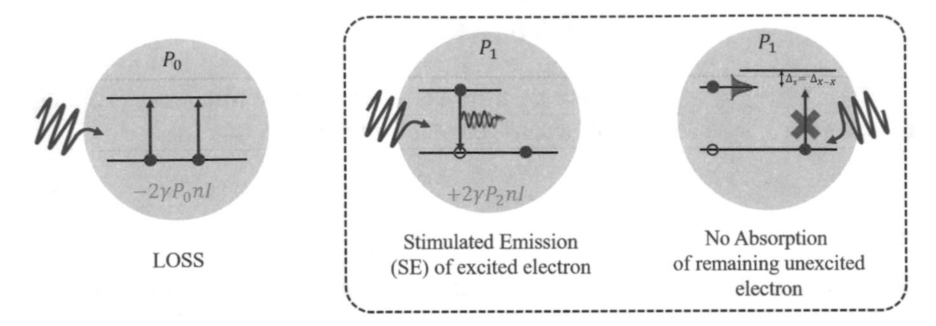

Fig. 3.16 Schematic of single exciton gain engineered in type II CdS/ZnSe core–shell QDs. The internal electric field generated towards the CdS core causes a transient stark's shift with magnitude $\Delta_s = \Delta_{X-X}$ equal to the repulsive coulombic energy as a consequence of charge localisation

where γ and n_c are the excitonic recombination probability and number density. The first term corresponds to absorption loss of unexcited QDs while the second term corresponds to stimulated emission contributions from singly-excited QDs in response to the transient stark's effect. The single exciton gain threshold can be obtained by solving Eq. (3.22), where substituting $P_0 = 1 - P_1$ and letting $\Delta I = 0$ gives us the result $P_{1,\text{th(X)}} = \frac{2}{3}$.

Assuming ideal conditions, where the transient stark's shift is larger than the linear absorption linewidth ($\Delta_s > \Gamma$), this tells us that minimally two-thirds of the excited ensemble must be singly excited, which translates to the threshold condition:

$$\langle N \rangle_{\text{th(X)}} = 0.67 \tag{3.23}$$

In this case, $P_{1,\text{th(X)}} \equiv \langle N \rangle_{\text{th(X)}}$ because we considered the gain contribution to only come from singly-excited QDs and ignoring higher order contributions. As such, this approximation fundamentally defies the Poisson statistics describing photoexcitation in the ensemble. In less ideal situations, where $\Delta_s \sim \Gamma$, spectral overlap between absorption and stimulated emission will arbitrarily raise the threshold condition, such that it can be generalised as a function of (Δ_s/Γ) as follows:

$$\langle N \rangle_{\text{th(X)}} = \frac{2}{3 - e^{-(\Delta_s/\Gamma)^2}} \tag{3.24}$$

where $e^{-(\Delta_s/\Gamma)^2}$ is the relative Stark's shifted detuning term. Note that it is important for $\Delta_s > 0$ to be repulsive in nature, as an attractive potential would not only promote the formation of higher-ordered excitons but also further raise the single exciton gain threshold. Readers interested in strategies used in the development of single exciton gain in Chalcogenide QD systems are encouraged to refer to various literatures [20, 49, 50].

Although fabricating core–shell structured PNCs for similar studies are still in developmental stages [51], Zhao et al. was the first to report single exciton gain in $CsPbBr_3$ PNCs embedded into Au–Si core/porous shell nanocomposite systems [52]. From Fig. 3.17a, optical build-up of excitons in the nanocomposite is engineered through resonance energy-transfer (RET) mechanisms, where thermalising hot electrons from $CsPbBr_3$ PNCs are transferred to a local surface plasmonic resonance (LSPR) level created by the innermost Au-core [52]. Figure 3.17b shows the TA spectra of the nanocomposite system pumped at 12 $mJcm^{-2}$, where excited state absorption (ESA), PB and stimulated emission (SE) are assigned to the 480 nm, 520 nm and 545 nm probed bands, respectively [52]. Due to early-time spectral overlaps between the 520 nm-PB and 545 nm-SE bands, a four-peak fit deconvolution was performed and shown in Fig. 3.17c.

In doing so, the relation of SE position with probe delay time can be plotted (Fig. 3.17d) and fitted monoexponentially to give ~340 fs. Given that this value is consistent with LSPR relaxation lifetimes, and since the gain build-up is limited by

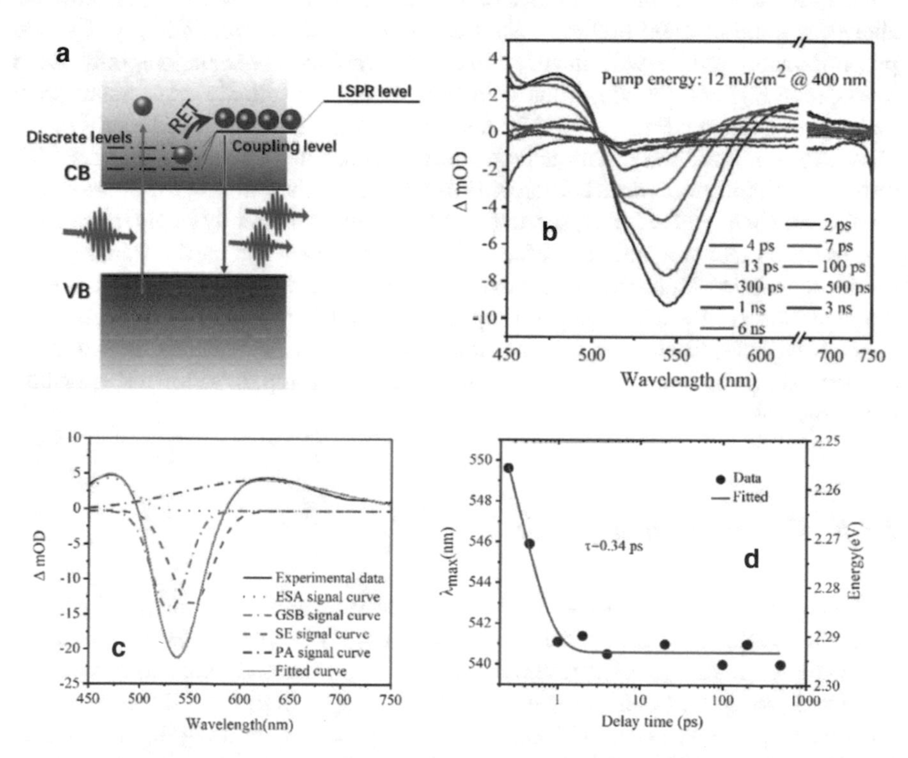

Fig. 3.17 Single exciton optical gain in Au-porous Si injected with $CsPbBr_3$ PNCs (nanocomposite) [52]. **a** Schematic of the excitonic optical buildup via resonant energy transfer (RET) from hot electronic states to a relatively higher local surface plasmon resonant (LSPR) metastable state. **b** TA spectra of the nanocomposite system pumped at 12 $mJcm^{-2}$ in optical gain regime. **c** Deconvolution of a TA spectra resolved at early delay time of 1 ps. **d** Relationship between stimulated emission position as a function of probe delay time

RET to the LSPR state, the 340 fs component was assigned to the excitonic gain build-up lifetime and is much faster than typical carrier-phonon interactions [52]. TA kinetics spectrally resolved in the SE band also revealed an excitonic gain decay lifetime of ~146 ps [52].

Single exciton gain and lasing was also reported in several unengineered systems of $Cs_{0.17}FA_{0.83}PbBr_3$ PTFs [53], $CsPbBr_3$ PNC films [54] and $CH_3NH_3PbBr_3$ PTFs [55]. In $Cs_{0.17}FA_{0.83}PbBr_3$ PTFs, single exciton gain was concluded from TA kinetics on the basis of low threshold carrier densities ~6.51×10^{17} cm^{-3} in the absence of auger recombination [53]. In addition, they showed that inelastic Frohlich scattering of excitons with LO phonons with ($\hbar\omega_{LO} \sim 29$ meV) is insufficient to dissociate them into free-carriers. However, the role of mixed cations and its carrier/energy dynamics leading to excitonic build-up and the origins of repulsive exciton-exciton interactions was not mentioned and remains elusive. Shortly, Juan et al. also claimed to observe single exciton gain in both $CsPbBr_3$ and $CsPbI_3$ PNC films [54]. Here, the basis of conclusion was that the red-shifted ASE is a consequence of self-absorption effects that depended on film thickness and that at $1 < \langle N \rangle < 2$, TR-PL kinetics showed a trend of fitted lifetime τ shortening with fluence, which imply Φ_f also proportionately reduce with fluence [54]. However, such observations only seem to disprove higher-ordered exciton gain but not prove excitonic gain, which may invite strong debate. Firstly, the ASE's red-shift is likely a combined effect of re-absorption minimisation and attractive exciton-exciton interactions. Thus, lesser ASE red-shift in thinner films does not contain information on the nature of exciton-exciton interaction energies. Secondly, shortening of τ should not directly imply decreasing Φ_f, as the former and latter are affected by fluence dependent carrier trapping and carrier dynamical processes, respectively. Instead, a direct measurement tracking the evolution of Φ_f with fluence in optical gain regimes would have been more conclusive. Thus, future works in this area should focus on elucidating the nature of exciton-exciton interactions and its role in excitonic build-up for different Perovskite morphologies.

3.1.6 Polariton Optical Gain

In Sects. 3.1.2, 3.1.3, 3.1.4 and 3.1.5, we have discussed a wide range of photon lasing mechanisms, where optical build-ups are largely dictated by carrier dynamics and energetic landscapes of Perovskites and carrier-photon coupling was completely neglected [56, 57]. For anisotropic systems offering (I) high E_B^X, (II) high exciton effective masses [58, 59] and (III) high oscillator strengths, exciton-photon coupling becomes non-negligible and new quasiparticles called polaritons can manifest in the excited system. Importantly, exciton-photon coupling strength depends greatly on the material's oscillator strength which would affect the polaritonic stability. Interestingly, the polaritons possess half-light half-matter properties, allowing the generation of nonclassical light output for laser-based and cavity quantum electrodynamical applications [60]. One example is polariton lasing, where optically coherent light is emitted as a result of macroscopic polaritonic condensates (PCs) [61], analogous to the conventional Bose–Einstein Condensate (BEC) in ultracooled atoms [62]. Since PCs act without having to satisfy population inversion, it is deemed more pump-efficient than photon lasing [60]. Generally, PC formation depends on the magnitudes of Rabi-splitting energy Ω and the exciton-photon detuning energy Δ. Here, Ω denotes the coupling strength and is defined as the energy difference between the UP and LP bands at normal incidence $k_{||} = 0$. Conventionally, this coupling is well described by the coupled-oscillator model, where energetic anti-crossing between them gives rise to an upper (UP) and lower (LP) polariton branch. The UP band is usually not observed in experiments, as it is outcompeted by electron–hole continuum absorption, thermal relaxation, and the high reflectance from the top DBR [61]. As such, polariton lasing occurs mostly in the LP branch [61]. On the other hand, Δ denotes the dominant particle of the coupling expressed as the difference between their energies [63]:

$$\Delta = E_C - E_X \tag{3.25}$$

where E_C denotes the cavity photon energy. If $\Delta > 0$, the polariton is "exciton-like", which possess long recombination lifetimes and is susceptible to acoustic phonon interactions. Likewise, $\Delta < 0$ suggests the polariton to be "photon-like" and are susceptible to inefficient scattering with acoustic phonons. The polariton's exciton fraction $|X|^2$ can be calculated if Ω and all dispersion curve energies are known [63]:

$$|X|^2 = \frac{\Omega^2}{\Omega^2 + (E_X - E_{LP})^2} \tag{3.26}$$

where $E_X - E_{LP}$ is the energy difference between the exciton and LP branch at $k_{||} = 0$. For polariton lasing, polaritons with photon-like behaviours are preferred, as scattering with acoustic phonons are inhibited (polariton relaxation bottleneck) [64] and allows for stimulated scattering of LPs down to its PC build-up. After that, spontaneous radiative decay of the polaritons in the condensate produces coherent emission of polariton lasing [65]. In other words, we require the system to possess large Ω and $\Delta < 0$ ($|X|^2 < 0.5$) under non-resonant excitation conditions [61, 66]. From Eqs. (3.25) and (1.17), $\Delta < 0$ can be achieved by reducing E_C via increasing cavity lengths [67].

In the past, polariton lasing in inorganic systems [68–71] was only observed in cryogenic temperatures due to weak Wannier-mott $\Delta_X < 10\,\text{meV}$ and mostly met with lattice mismatch and thermal strain issues during integration into cavities. On the other hand, organic materials possessing intrinsic disorderliness and defects have been found to negatively impacts the PC process [72–74]. From the robust optical properties and defect tolerance of Perovskite materials, we expect Perovskite polariton lasers to hold great potential and are currently in developmental stages [57, 61, 75–77]. In 2017, Su et al. was the first to demonstrate room temperature polariton lasing in CVD-grown $CsPbCl_3$ PMPLs ($\sim 12\,\mu m$) based VCSELs [61].

Large $\Omega \sim 265$meV and $\Delta \sim -25$ meV$\left(|X|^2 \sim 0.45\right)$ indicative of strong coupling and photon-dominant polariton behaviours were reported in these PMPLs, which are ideal for PC formation and polariton lasing. In characterising polariton gain, it is important to verify the existence of PC. Thus, observing a clear crossover from uncondensed polaritons to PCs from below to above threshold pumping fluences is necessary. The onset of condensation occurs whenever the average separation between polaritons are comparable or lesser than its De-Broglie wavelength, causing coherent overlap of polaritonic wavefunctions [78]. As such, stimulated scattering occurs where macroscopic polaritonic relaxations governed by Bose–Einstein distribution [78] occurs, transiting them down to its lowest energy quantum state [79]. Figure 3.18a, b show the angle-resolved LP emissions before and after its condensate threshold [61].

Here, the $CsPbCl_3$ PMPLs show the narrowing of LP emission towards the normal direction $k_{||} = 0$ and a slight blueshift above the LP dispersion curve when pumped above PC threshold. This observation is a result of macroscopic stimulated scattering of "uncondensed" photon-like polaritons in the system, causing them to subsequently occupy the lowest energy polaritonic ground-states. Figure 3.19c shows the fluence dependent PL spectra and Fig. 3.19d shows the simultaneous FWHM narrowing and nonlinear increase in integrated light intensity, with a clear threshold at 12 μJcm^{-2}. Clearly, these two features are similar to photonic gain and are indicative of optical

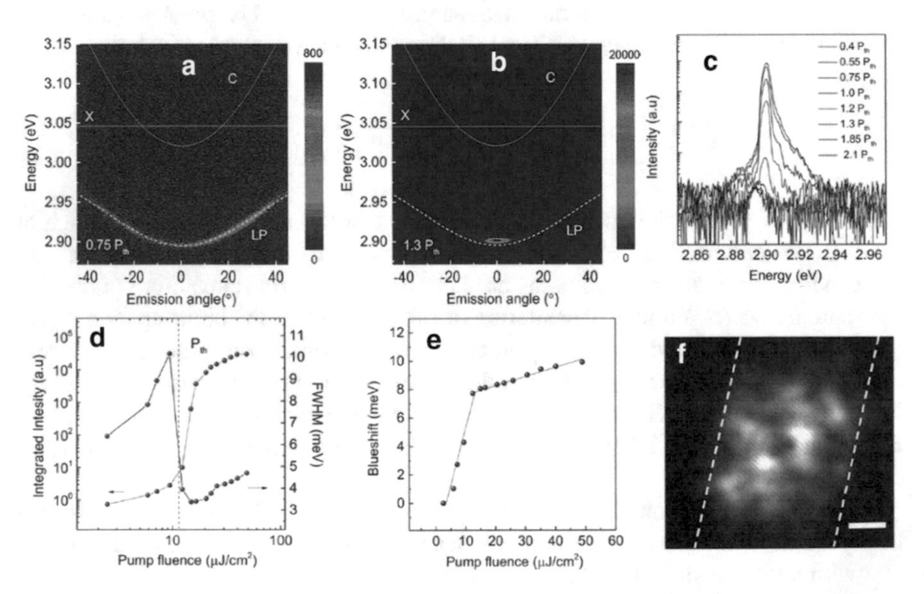

Fig. 3.18 Polaritonic condensation (PC) in $CsPbCl_3$ PMPLs [61]. Angle-resolved PL mapping taken at **a** 0.75Pth and **b** 1.30Pth. **c** Fluence dependent PL spectra extracted at $k_{||} = 0$. **d** Nonlinear growth of integrated intensities, FWHM narrowing and **e** Fitting of experimental (blue circles) blue-shifting energy with its theoretically calculation (red line) at $k_{||} = 0$ LP branch are evidence of PC. **f** Interference pattern acquired from summing real and inverted Centrosymmetric space PL Images. Scale bar: 4 μm

coherence of polaritonic lasing. Note that the subsequent broadening of LP FWHM with fluence at above threshold is due to decoherence induced by active polariton-polariton interactions [80] and possibly other localised PC modes [81]. Figure 3.18e shows the close agreement between experimentally obtained (blue circles) LP blue-shift and the theoretical model (red fitting line) derived from coupling the driven-dissipative Gross-Pitaevskii equation to the rate equation of the exciton reservoir [61]. At below PC threshold, the LP blue-shift is larger (steeper) due to increasing exciton–polariton interactions mediated by increasing fluences. However, the blue-shift generally tapers as the interaction dynamics is now dominated by polariton-exciton interactions [61]. Finally, Fig. 3.18f shows the superimposed image between the real and inverted Centro-symmetrical PL images acquired from interferometry [61]. Importantly, checking for the long-range spatial coherence via interferometry is a well-established technique in confirming PC and hence polariton lasing. Here, the bright spots observed are associated with the localisation of PCs, possibly due to photonic disorder during fabrication of $CsPbCl_3$ PMPLs [61]. Importantly, the superposition between the two PL images produce interference fringes that are observable over $\sim$ 15 μm, which indicate long-range spatial coherence associated with the formation of PCs [61]. Observation of Polariton lasing has also been reported in $CsPbBr_3$ [82–84] and $CH_3NH_3PbBr_3$ [85] nanostructures as well as two-dimensional Perovskites [86, 87].

3.2 Fabrication Techniques for Perovskite Lasers

In this section, we present several approaches to fabricate high quality natural and external cavities for low pumping threshold Perovskite lasing. In recent years, nanoimprinting, lithographic and inkjet printing are gaining attention due to their potential in realising large-area laser arrays towards upscaling and commercialisation. Importantly, we also focus on strategies proposed for device thermal management in realising CW pumped Perovskite lasing, which is an important milestone towards realising electrically pumped Perovskite laser devices. For clarity, we will discuss key optical characterisations for each cavity configuration that ascertains lasing.

3.2.1 Fabry–Perot Perovskite Lasers

Optically pumped Perovskite lasers embrace the VCSEL configuration, where the active Perovskite layer is sandwiched between two DBR mirrors, as shown in Fig. 3.19. Sometimes, a transparent spacer such as PMMA is spincoated on top of the active layer to provide surface passivation before another DBR is fixed on top of the device [11]. VCSELs are desirable because its laser output is strongly directional and its circular beam profile directed perpendicular to its device surface makes it easy

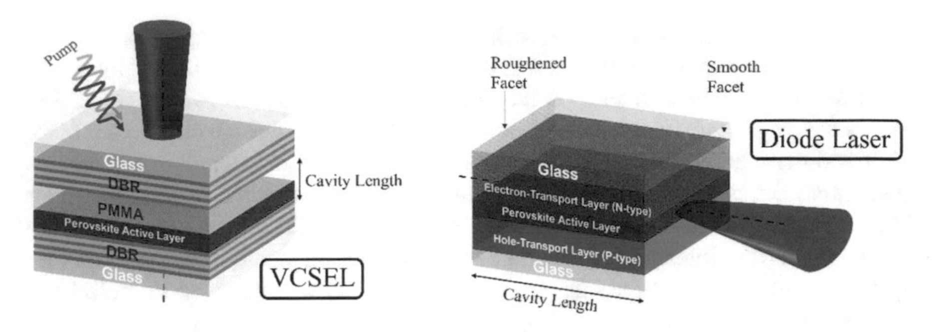

Fig. 3.19 A schematic showing the conceptual difference between a VCSEL and diode lasers of the Fabry–Perot configuration

to couple with optical fibers for optical communications [68]. Other applications of VCSEL structures include high density optical storage, laser displays, parallel optical computing, and signal processing [88, 89]. On the other hand, Perovskite diode lasers could also be less appealing than Perovskite VCSELs as the former produces elliptical and diverging side-emitting outputs [90]. From a manufacturing standpoint, diode lasers also suffer from additional dicing costs during quality-control testing, as its side-emitting nature meant that a full wafer must be diced before a single chip can be mounted.

Perovskite VCSELs are easy to integrate because it does not suffer from stringent band-alignment [91] and lattice mismatch conditions [92, 93]. Spectroscopic verification of Fabry–Perot etalon resulting in cavity modes was investigated by Wang et al. in 2017 [90]. Here, maximum optical feedback was ensured by situating the PNC's PL band within its DBR near-unity "stopband", with the inset illustrating the PNC-based VCSEL device configuration (Fig. 3.20a). At below lasing thresholds, a series of sharp "spikes" was ascribed to Fabry–Perot cavity modes because (i) the mode number assignment corroborates with Eq. (1.21) and (ii) similar features were also observed in low intensity CW excitations [90]. Wang et al. reasoned that the cavity modal envelop is red-shifted from the PNC PL band due to reabsorption effects caused by radiation multi-pass during round-trip feedbacks [94]. Next, angular dependent steady-state PL was employed to study the cavity mode behaviour [90]. As shown in Fig. 3.20b, all cavity modes blue-shift with increasing detection angle θ. As shown in Fig. 3.20c, this occurs because off-axial mode (black line, with angle ϕ) generally propagate longer distances. Thus, their contribution to lasing intensities at detection angle θ can be derived by considering Snell's law of refraction at interfaces:

$$n_p \sin\phi = \sin\theta \tag{3.27}$$

where n_p is the Perovskite's refractive index and $L = L_{eff}\cos\phi$. By combining Eqs. (1.21) and (3.27), the relation between mode λ_m and θ is given by [90]:

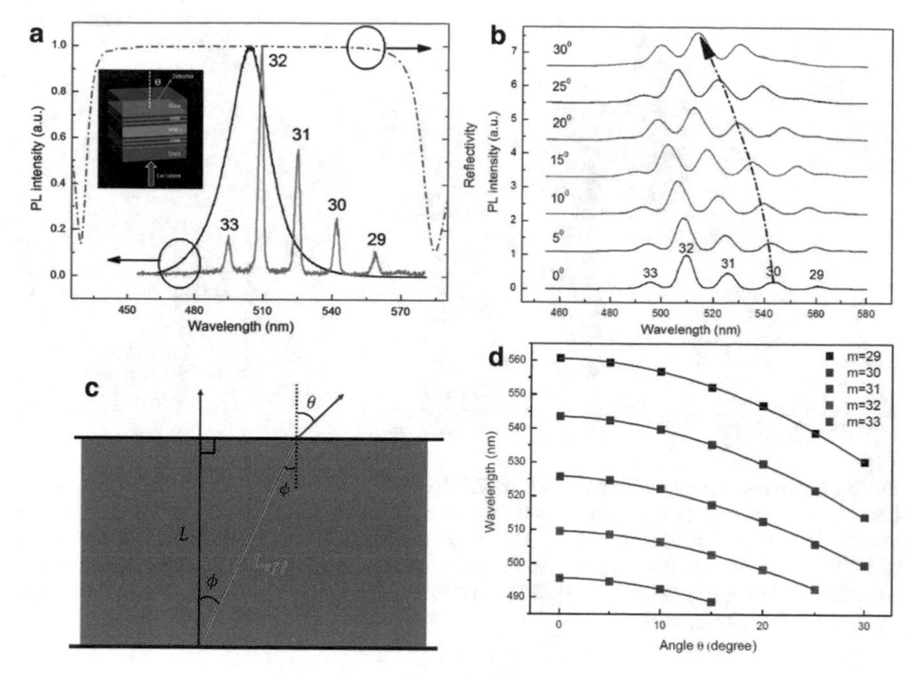

Fig. 3.20 Fabry–Perot cavity mode characterisation in $CsPbBr_3$ PNC-VCSELs [90]. **a** An overview of PNC ensemble PL, VCSEL cavity mode relative to the DBR reflectivity spectra overlaid together. **b** Steady-state PL spectra of the VCSEL with detection angle θ and **c** verifying Fabry–Perot mode behaviour with θ fitted with Eq. (3.28). **d** Illustration of off-axis mode oscillations responsible for angular-dependent blueshifts

$$m\lambda_m = 2L_{eff}\sqrt{n_p^2 - \sin^2\theta} \quad (3.28)$$

Since Eq. (3.28) is seen to fit the blue-shifting trends of cavity modes with θ in Fig. 3.20d, the cavity modes are assigned as Fabry–Perot cavity modes. Figure 3.21 shows a set of Fabry–Perot lasing characterisation conducted by Wang et al. in in the $CsPbBr_3$ PNC based VCSEL. From Fig. 3.21a, an increase in ~fs pulsed excitation fluence led to the development of a single mode lasing operation along m = 32. Figure 3.21b, c shows the simultaneous m = 32 cavity mode FWHM narrowing and non-linear mode intensity growth once pumped above lasing threshold of ~11 μJcm^{-2}, respectively. Furthermore, the bottom inset of Fig. 3.21a showed that them = 32 lasing output is strongly collimated, with divergence of $\sim 3.6°$, as compared to the PNC's PL (above inset) at $\sim 15°$ [90]. Ease of PNC integration onto DBRs with suitable stopbands of blue- and red-emitting PNC VCSELs have also been demonstrated by Wang et al. and shown in Figs. 3.21d, e.

Interestingly, the report also demonstrated the effects of active layer thickness, where the thicker layered VCSEL produced multimode lasing, albeit with significantly lower threshold at $9\mu Jcm^{-2}$ than the single mode VCSEL (~900

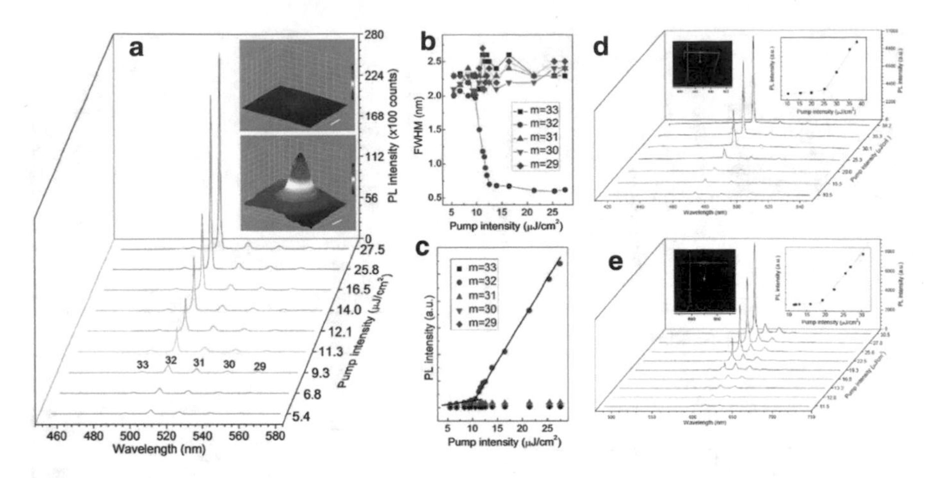

Fig. 3.21 Lasing characterisation in $CsPbBr_3$ PNC-VCSEL [90]. **a** Pump fluence dependent evolution of cavity modes. Inset: Output beam profiles at below (upper) and above (bottom), lasing thresholds, with divergences of ~15° to 3.6°, respectively. Scale bar: 1 mm. **b** Further narrowing of FWHM from 2 nm to 0.6 nm and **c** Non-linear intensity growth in m = 32 lasing mode beyond 11 μJcm^{-2}. Demonstration of PNC-VCSELs in **d** blue and **e** red lasing outputs with $CsPb(Cl/Br)_3$ and $CsPb(Br/I)_3$ PNCs, respectively

μJcm^{-2}),under ~ ns pulsed excitation conditions [90]. Another important characterisation is the VCSEL's output polarization. Strong linearly-polarized outputs are desired in interferometry, optical modulations, and non-linear frequency generations (where phase-matching is required) [95, 96]. The degree of polarization (DOP) of a laser wave is defined as [97]:

$$\text{DOP} = \frac{I_{max} - I_{min}}{I_{max} + I_{min}} \tag{3.29}$$

where I_{max} and I_{min} are the maximum and minimum output intensities at orthogonal analyser detection angle ζ. Essentially, the relation of lasing intensity with analyser angle ζ is given by Malus' law [96]:

$$I = I_0 \cos^2 \zeta \tag{3.30}$$

In ideally linear-polarised outputs, $I_{min} = 0$ when $\zeta = 90°, 270°$, which gives DOP = 1. From Fig. 3.22c f, DOP values of ~ 0.40 and 0.81 were determined in $CsPbBr_3$ PNC-based and $CH_3NH_3PbI_3$ PTF-based VCSELs. Partially linear polarised outputs from $CsPbBr_3$ PNC-based VCSELs (Fig. 3.22a, b) were attributed to the competition between intrinsically polarised excitonic emissions between three-fold degenerate bright-triplet and a dark singlet excitonic transitions [96]. On the other hand, strong linearly polarised output from $CH_3NH_3PbI_3$ PTF-based VCSELs

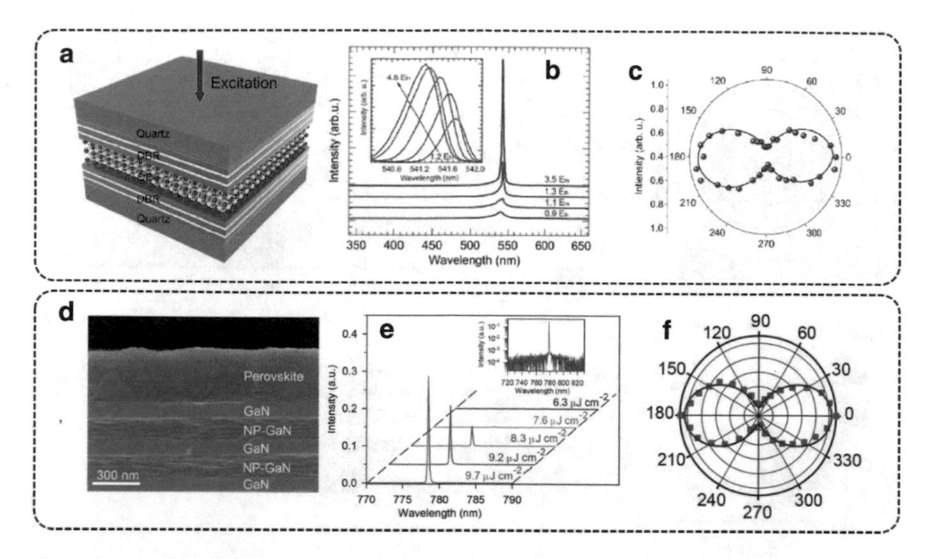

Fig. 3.22 **a–c** Schematic, fluence dependent lasing emissions and normalised lasing intensity with respect to polarization angle ϕ of a $CsPbBr_3$ PNC-based VSCEL, respectively [96]. **d–f** SEM image, fluence dependent lasing emissions and normalised lasing intensity with respect to polarization angle ϕ of a $CH_3NH_3PbI_3$ PTF-based VSCEL, respectively [98]

(Fig. 3.22d, e) was attributed to optical anisotropy arising from film thickness inhomogeneity, residual strain in the DBR and/or nanopore alignment [98]. Interestingly, lattice-matched GaN(intrinsic)-GaN(n-doped) layer pairs grown epitaxially showed effective feedback coupling to the PTF active layer, resulting in Q ~ 1110 and ultralow lasing threshold ~7.6 μJcm^{-2} [98].

To date, successful CW-pumped lasing in the Fabry–Perot configuration has only been demonstrated in fabricated VCSELs and PNWs [83], specifically under strong exciton-photon coupling circumstances. In a recent report, Shang et al. demonstrated CW-pumped polariton lasing with ultralow threshold at $P_{th} \sim 130$Wcm^{-2} in $CsPbBr_3$ PNWs (height H ~180 nm) under 7.8 K conditions (Fig. 3.23e) [83]. The success was credited to two strategies taken to facilitate efficient thermal diffusion along the PNW-substrate interface, as shown in Fig. 3.23a. The first strategy involves studying the effects of width-to-height (W/H) ratio in the PNW-substrate interfacial heat transfer, as shown in Fig. 3.23b. He observed that PNWs with $H > 300$ nm are too thick and hinders thermal transfer while PNWs with $H < 50$ nm are too thin and photonic mode localisation is impaired by optical diffraction limit [83]. The second strategy taken was to select the best thermal conductive substrate after balancing between substrate thermal conductance and dielectric function. Figure 3.23c shows the simulated transient temperature of PNWs coupled to four different substrates: Sapphire $\left(42\,\text{Wm}^{-1}\text{K}^{-1}\right)$, Ag $\left(429\,\text{Wm}^{-1}\text{K}^{-1}\right)$, Glass $\left(1.4\,\text{Wm}^{-1}\text{K}^{-1}\right)$ and Si $\left(148\,\text{Wm}^{-1}\text{K}^{-1}\right)$.

While high substrate thermal conductance can lower PNWs' stabilising temperatures, Fig. 3.24d also revealed that candidates like Ag and Si possess relatively large

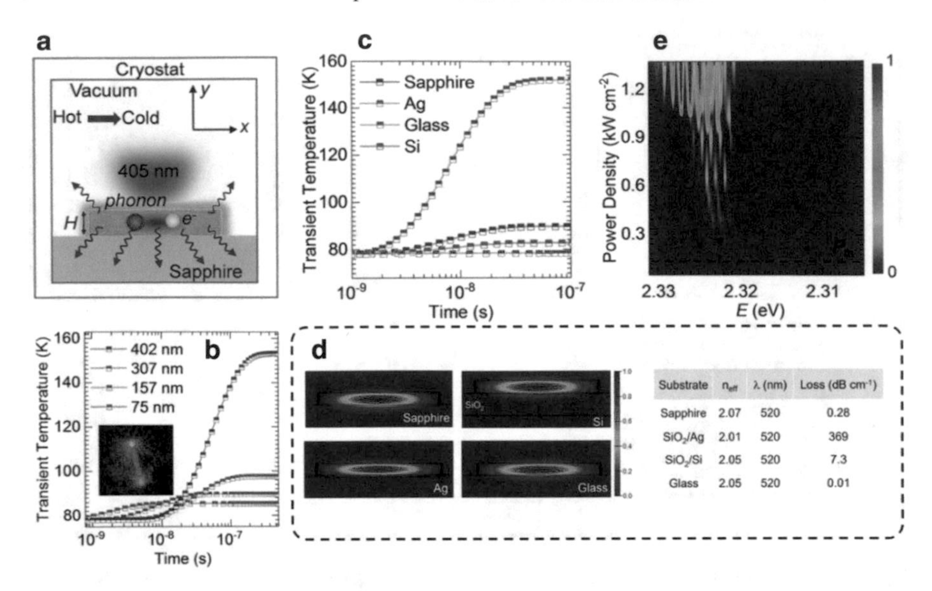

Fig. 3.23 Factors leading to successful CW-pumped $CsPbBr_3$ PNWs at 7.8 K, with lasing threshold of $\sim$ 130 $Wm^{-1}K^{-1}$[83]. **a** Illustration of height-dependent PNW-sapphire interfacial thermal diffusion in cryostat environment. Simulated **b** height-dependent PNW transient temperature and **c** substrate-dependent PNW (H = 160 nm) transient temperature at P = 1.75 $kWcm^{-2}$ via DEVICE (Lumerical Inc.). **d** Simulated modal distributions in PNWs (H = 400 nm) via MODE (Lumerical Inc.) and the substrates' corresponding dielectric function-induced optical losses. **e** Two-dimensional pseudocolored PL emission in a H = 180 nm PNW thermally dissipated by a sapphire substrate under 7.8 K conditions

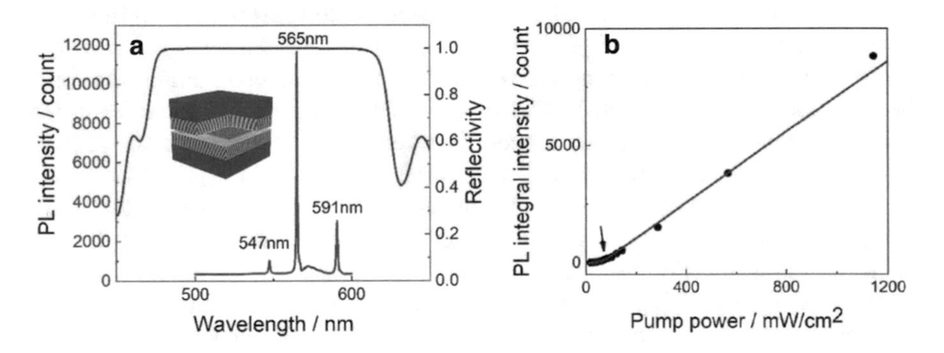

Fig. 3.24 Room temperature CW-pumped lasing in $CH_3NH_3PbBr_3$ PSC-TF based VCSELs [100]. **a** A comparison between the DBR's stopband and the VCSEL's output when pumped at ~1200 $mWcm^{-2}$. **b** Non-linear intensity growth of the 565 nm mode beyond 34 $mWcm^{-2}$, indicating laser mode oscillations in the VCSEL

dielectric functions that induce significant radiation leakages along the interface, thus increasing lasing thresholds [83]. Hence, the balance between thermal conductance and dielectric function induced loss is best found in sapphire substrates. CW-pumped Polariton lasing in $CsPbBr_3$ PNWs have also been reported by Evans et al., which includes a comprehensive set of characterisations [77]. On the other hand, room-temperature enhanced ASE in CW-pumped $CH_3NH_3PbBr_3$ PTF-based VCSELs was first reported by Alias et al., with a high threshold of ~89 kWCm^{-2} [99]. Here, the active layer is sandwiched between two PMMA protective layers to foster surface passivation and improved waveguiding effects necessary for minimising losses. However, the output's lack of DOP and relatively large narrowband FWHM puts it under the debate of whether lasing was observed. Meanwhile in 2019, room temperature CW-pumped lasing in $CH_3NH_3PbBr_3$ SC-TF based VCSELs was reported by Cheng et al., as shown in Fig. 3.24 [100]. Figure 3.24a shows the lasing spectrum, which features a multi-mode lasing output with FWHM ~0.8 nm, under pumping density of ~1200 mWcm^{-2} [100]. Interestingly, the successful demonstration of room-temperature CW-pumped lasing here did not account for thermal management and is simply assigned to strong exciton-photon interactions with estimated Rabi splitting energy ~372 meV [100].

In practice, a VCSEL's output consistency greatly depends on the Perovskite active layer thickness, uniformity, and cavity length. Since single-mode lasers imposes an upper limit of Perovskite layer thickness, one must pay extra attention to the film's surface roughness. Another key factor is to minimise the trapped air columns when the top DBR mirror is brought to contact with the active layer and bottom DBR mirror. In some reports, effective cavity length L have been estimated ~tens of μm [90] and are much thicker than expected active layer thicknesses. This is likely due to trapped air columns as shown in Fig. 3.25a. Occasionally, discrepancies in expected and calculated FSR can also possibly stem from involvement of higher-ordered transverse waveguided modes in addition to fundamental TE_{00} and/or TM_{00} modes. A set of extensive discussion in Fabry_Perot lasing in $CH_3NH_3PbBr_3$ PMWs is presented by Gu et al. [101]. Experimental discrepancies between expected and calculated FSRs often stem from the Next, a possible cause of irregular VCSEL

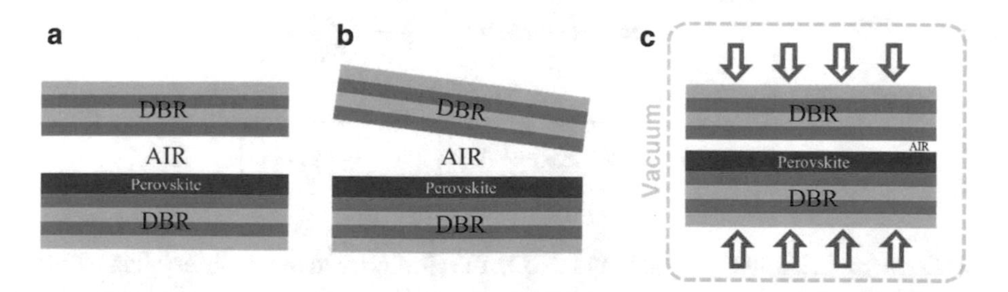

Fig. 3.25 Overview of common VCSEL fabrication mistakes. **a** Air columns are trapped in ambient fabrication. **b** Inhomogeneous adhesion leads to uneven cavity lengths that affects lasing output reproducibility. **c** By applying homogeneous force during adhesion of top DBR under vacuum environments, air columns can be minimised

output in terms of lasing position and thresholds is likely due to inhomogeneous adhesion or huge film surface roughness, where a visually indistinguishable angular tilt causes a cavity length gradient, as shown in Fig. 3.25b. Thus, air columns and angular tilts in VCSEL fabrication can be minimised by performing the top DBR adhesion in vacuum-purged environment and providing homogeneous force (such as via a G-clamp), as shown in Fig. 3.25c.

3.2.2 Whispering Gallery Mode (WGM) Perovskite Lasers

In Sect. 2.2, we discussed that the PMPLs and PMCs are excellent morphological systems that offers WGM lasing without having to fabricate any external cavity. Generally, WGM modes localises through resonant recirculation of constructive interfering light waves via total internal reflection (TIR). Thus, this low-loss optical feedback mechanism allows WGM to possess extremely narrow lasing linewidths with extremely high Q-factors. WGM lasing is easy to achieve because its refractive index (~2) is higher than its surrounding air [102–104]. Due to its sensitivity towards the material's polygonal shape, Perovskite WGM lasers can find potential application as optical sensors. Optically pumped WGM lasing was first observed in $CH_3NH_3PbI_3$ single-crystalline PMCs [19] and subsequently in capillary-filled PNCs [2, 105, 106] and other PMPLs [107–110] systems. WGM lasing from Silica microspheres conformally coated with $CH_3NH_3PbI_3$ via atomic layer deposition (ALD) [111] and with as-synthesized $CsPbBr_3$ PNCs [112] were also reported as alternative methods.

Generally, WGM modes are characterised by transverse electric (TE) or magnetic (TM), radial order q and azimuthal order m. Mostly, only the first radial order ($q = 1$) fundamental is observed due to highest Q-factor that is dependent directly on the system's polygonal geometry [113, 114]. The first radial ordered TE and TM WGM modal positions are solved from Lam et al.'s asymptotical function and is given as [115]:

$$\lambda_{m(TE)}^{(1)} = \frac{\pi n_{cav} D_{cav}}{m + 1.856m^{1/3} + \left[\frac{1}{2} - \frac{n_{cav}}{\sqrt{n_{cav}^2 - 1}}\right]} \tag{3.31}$$

$$\lambda_{m(TM)}^{(1)} = \frac{\pi n_{cav} D_{cav}}{m + 1.856m^{1/3} + \left[\frac{1}{2} - \frac{n_{env}}{n_{cav}\sqrt{n_{cav}^2 - 1}}\right]} \tag{3.32}$$

where $m \gg \frac{1}{2}$. From Eqs. (3.31) and (3.32), the similarity in their solution just before the parenthesis in their denominator tells us that $\lambda_m^{(1)} \sim \frac{\pi n_{cav} D_{cav}}{m+1.856m^{1/3}}$ is the "reference" mth azimuthal ordered WGM mode comprising of a TE-TM pair distinguished by their terms in parentheses. The corresponding WGM FSR for respective TE and TM modes is calculated to be:

$$FSR^{(1)}_{\lambda(WGM)} \approx \frac{\left(\lambda^{(1)}_{m}\right)^2}{\pi n_{cav} D_{cav}} \tag{3.33}$$

where $FSR^{(1)}_{\lambda(WGM)} \propto D^{-1}_{cav}$ and is similar to Fabry–Perot lasing FSR in Eq. (1.19). The net Q-factor provided by a Perovskite microstructure is governed by its radiative waveguide losses, material reabsorption, surface scattering and absorption losses, respectively. This relation of WGM Q-factor and losses are related by the following inverse law [115]:

$$\frac{1}{Q_{net}} \approx \frac{1}{Q_{rad}} + \frac{1}{Q_{mr}} + \frac{1}{Q_{ss}} + \frac{1}{Q_{sa}} \tag{3.34}$$

Here, Q_{rad} is the Q-factor that accounts for imperfect TIR feedback that causes light to stray out of the feedback loop. Lower azimuthal ordered modes are lossy as they take shorter round-trips with small reflection angles (sharp turns) and TIR conditions may not be met. Q_{mr} is the Q-factor that accounts for material reabsorption losses and is explicitly given by.

$Q_{mat} = \frac{2\pi n_{cav}}{\alpha\lambda}$ in most literatures [116]. Q_{ss} is the Q-factor that accounts for surface Rayleigh scattering losses caused by material surface roughness and is given by $Q_{ss} = \frac{\lambda^2 D_{cav}}{2\pi B \langle S \rangle^2}$ [117], where $\langle S \rangle$ and B are the root-mean-square roughness and length of surface inhomogeneity. Lastly, Q_{sa} accounts for surface-induced absorption losses caused by surface contaminant/defects. Based on the argument of surface-volume ratio, we expect $Q_{net} \propto D_{cav}$. This result is similar to Fabry–Perot lasers, where the pursuit of single mode WGM output in smaller sized Perovskite microstructures are achieved at the cost of lower net Q-factor.

One key discussion in Perovskite microstructure is the origins of lasing [118, 119]. In the case of PMPLs, lasing can either arise from in-plane Fabry–Perot lasing mode or edge-reflecting WGM lasing modes. In a study by Qi et al., angle-resolved μ-PL Fourier imaging revealed that WGM lasing typically occurred universally across all edge lengths L at sufficiently high fluences while Fabry–Perot lasing are produced only in larger $CsPbBr_3$ PMPLs with $L > 13\mu m$ that would eventually also be suppressed at even higher fluences (see Fig. 3.26c) [118]. Figure 3.26a shows the fluence dependent integrated PL and lasing intensity on a $CsPbBr_3$ PMPL with $L \sim 13.7\ \mu m$. As fluence increases beyond the threshold of $\sim 31\ \mu Jcm^{-2}$, Fabry–Perot lasing is observed and is affirmed by the bright emission on two of its edges (Fig. 3.26b top inset). However, as pumping fluence exceeds $\sim 71\ \mu Jcm^{-2}$, the lasing intensity is seen to reduce and gradually pick up at ~90 μJcm^{-2}. Such observation was assigned to the competition and suppression of the initial Fabry–Perot lasing mode oscillation into WGM lasing mode recirculation (characterised by four corner bright emission, bottom inset of Fig. 3.26b). The transition from Fabry–Perot oscillation to WGM recirculation transition in different sized PMPLs was studied by observing the relation between lasing FSR and L via its group refractive index n_g. Here, the Fabry-Perot and WGM are distinguished by its slightly different values of n_g from the fitting

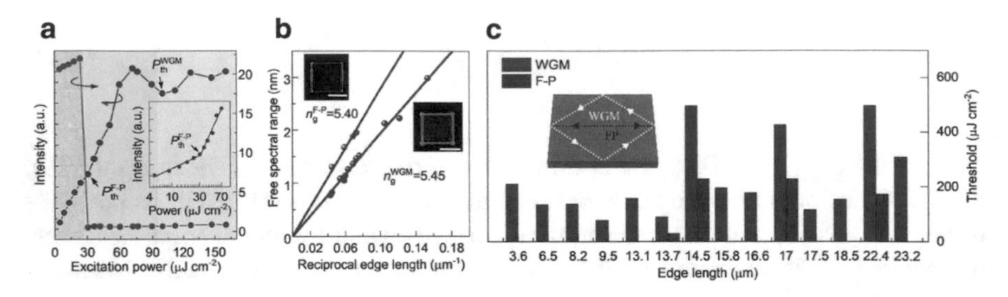

Fig. 3.26 Overview of Fabry–Perot and WGM lasing competition in $CsPbBr_3$ PMPLs [118]. **a** Fluence dependent integrated PL behaviour, showing a transition in lasing mechanism beyond 70 μJcm^{-2}. **b** Comparison of Fabry–Perot and WGM lasing mode FSRs across different PMPL edge lengths to extract their respective group refractive indices. **c** Statistical overview of Fabry–Perot and WGM mechanisms observed in terms of PMPL edge length and pump threshold

functions: $FSR_{FP} \sim \frac{\lambda^2}{2Ln_g}$ and $FSR_{WGM} \sim \frac{\lambda^2}{2\sqrt{2}Ln_g}$, respectively. Figure 3.26b showed that the fitted n_g for Fabry–Perot and WGM lasing was determined to be 5.40 and 5.45, respectively.

Amongst the many demonstrations of WGM lasing in Perovskite microstructures, frequency upconverted WGM lasing [120] with waterproof properties [121] at elevated operating temperatures, study of vernier-effect in near-proximity coupled Perovskite microspheres [122] are some key research highlights. Beyond the usual two-photon excitation, Shi et al. demonstrated three-photon (3PP) upconverted WGM lasing in vapor-phase epitaxial $CsPbBr_3$ PMPLs, with elevated operation temperature till 400 K [120]. As shown in Fig. 3.27a, the 3PP lasing threshold is remarkably low at $\sim 116\mu$Jcm^{-2} and has a record of sustained lasing under prolonged pulsed excitation for 10 h ($\sim 3.6 \times 10^7$ shots). Figure 3.27b shows the temperature dependent lasing

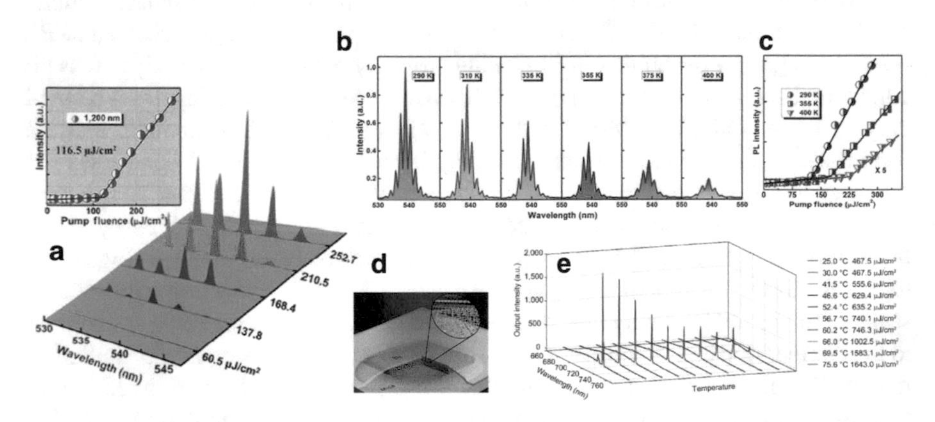

Fig. 3.27 **a** Frequency upconverted 3PP WGM lasing in $CsPbBr_3$ PMPLs, with **b** thermal tolerance up to 400 K, albeit with significant lasing intensity reduction and **c** increased lasing threshold [120]. **d** $CsPbI_3$ PMPL-hBN hybrid system for waterproofing and demonstrating **e** WGM lasing even with increasing temperatures [121]

spectra of a $CsPbBr_3$ PMPL with edge length 16.5 μm, ranging from 290 K up to 400 K under pumping fluence of $\sim$ 150μJcm^{-2}. Generally, lasing intensities degrade with increasing temperatures of operation are assigned to higher heat-induced non-radiative recombination rates [120] and could explain the increasing threshold trend with temperature in Fig. 3.27c. Successful 3PP WGM lasing with ultralow thresholds and elevated temperature operation were attributed to the successful growth of high quality PMPL resonators with enhanced oscillator strengths [120].

Figure 3.27d, e shows the $CsPbI_3$ PMPLs capped with hexagonal Boron Nitride (hBN) developed by Yu et al. with the intention of creating a water-proof and heat dissipating channel for the active $CsPbI_3$ PMPLs [121]. Evidently, the system also showed a similar lasing intensity degradation and increasing thresholds with temperature, as that from Shi et al.'s work [120]. In addition, the hBN encapsulation of $CsPbI_3$ PMPLs allowed stable WGM lasing in a myriad of polar solvents that include water, isopropanol, glycol and D38W for up to 24 h [121]. This observation is ascribed to the strong interfacial van-der-waals interaction between two inert hBN flakes that prevents polar molecules from invading the lasing PMPLs. Often, single-mode WGM lasers in smaller sized microcavities have much higher thresholds as more intense pumping is required to compensate for its poor modal confinement [120]. However, Zhou et al. showed that the vernier-effect coupling from differently sized Perovskite microspheres could force single-mode outputs with strong linear polarization properties without compromising modal confinements and thresholds [122]. As shown in Fig. 3.28a, b, when the two $CsPbBr_3$ microspheres gradually gets closer, the initially

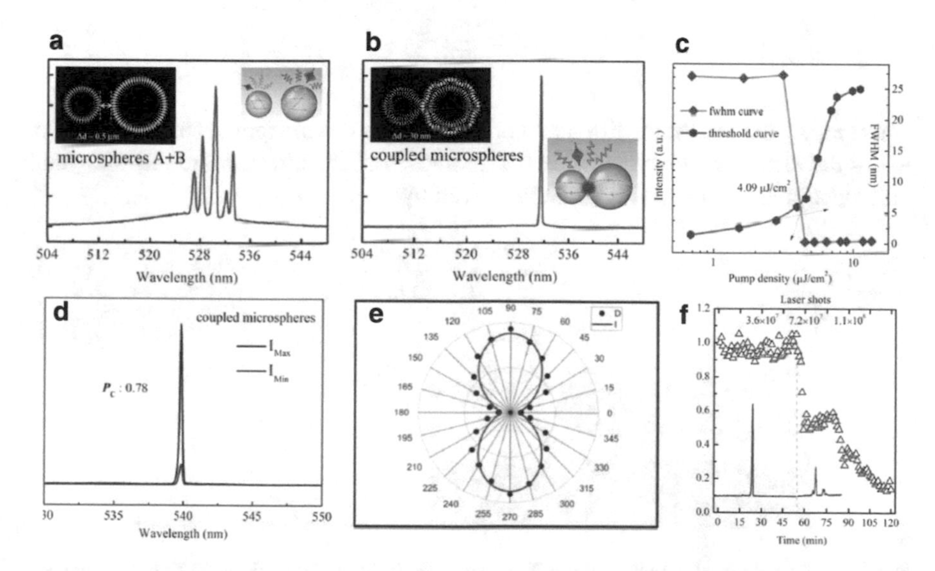

Fig. 3.28 Vernier-effect induced single-mode with strongly linear polarised WGM lasing emission [122]. Lasing spectra of **a** uncoupled microspheres (1.2 and 1.8 μm) at d = 0.5 μm and **b** coupled microspheres at d = 0.03 μm. **c** WGM lasing characterised by simultaneous FWHM narrowing and non-linear increase in light intensity. **d** Comparison of maximum and minimum lasing intensity **e** with respect to analyser angle, giving a strong DOP ~ 0.78. **f** Lasing stability sustained for ~55 min

uncoupled WGM emission becomes coupled, where the vernier-effect forces the only coinciding mode shared between both individual outputs to be remained and enhanced [122]. The right insets also shows that the presence of vernier coupling causes the coupled microspheres to share the same resonant recirculation pathway that drives highly polarised outputs. Figure 3.28c shows that the vernier-coupling gives rise to equally low lasing threshold $\sim 4\,\mu\text{Jcm}^{-2}$ and net Q-factor ~4100. Figure 3.28d, e shows the DOP characterisation of the vernier-coupled WGM output. Here, DOP is determined to be ~0.78 based on Eq. (3.28), which is a vast improvement from its uncoupled output at DOP ~ 0.2. Figure 3.28e shows that the prolonged excitation of vernier-coupled WGM lasing can sustain for ~55 min, implying that thermal management strategies need to be considered before implementing them for in on-chip photonics and optical sensing applications [122].

3.2.3 Distributed FeedBack (DFB) Perovskite Lasers

The Distributed Feedback (DFB) laser is an indispensable cavity configuration that offers reliable and spectrally narrow single-mode outputs, which can be applied in areas of high speed information networks and communications [123]. As shown in Fig. 3.29, the mode selectivity is handled by a reflective Bragg grating with a corrugation period Λ calculated by:

$$\Lambda = \frac{m\lambda_B}{2n_{\text{eff}}} \tag{3.35}$$

where m, λ_B are the Bragg diffraction order, Bragg's wavelength, respectively. Here, n_{eff} is the effective refractive index that results from the alternating Perovskite and grating/capping media and is explicitly given by:

$$n_{\text{eff}} = \frac{n_c w_c + n_p w_p}{w_c + w_p} \tag{3.36}$$

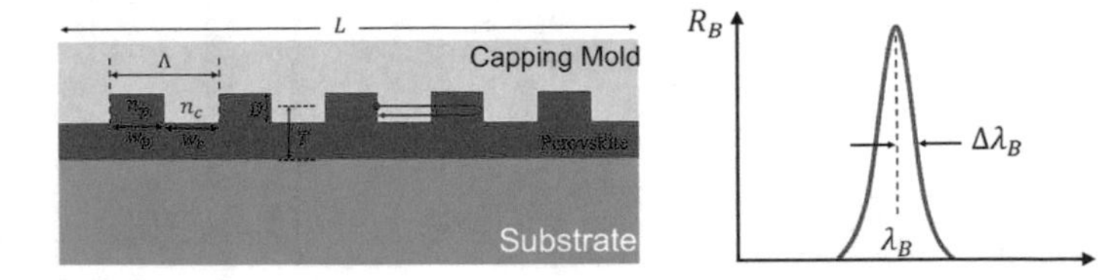

Fig. 3.29 Schematic of the DFB laser with mode selectivity determined by the Bragg reflectance spectrum R_B

where w_p, w_c, n_p and n_c are the Perovskite width, capping width, Perovskite index, and capping layer index, respectively. From Eq. (3.35), we require λ_B to be near the Perovskite layer's gain region for effective feedback and amplification and from Eq. (3.36), if $w_c = w_p$, then $n_{\text{eff}} = \frac{n_c+n_p}{2}$. In practice, single mode operation is achieved by minimising the Bragg reflectance R_B spectral width:

$$\Delta\lambda_B \approx \frac{\lambda_B}{Nm} \approx \frac{\lambda_B \Lambda}{mL} \approx \frac{\lambda_B^2}{2n_{\text{eff}}L} \tag{3.37}$$

where $N = \frac{L}{\Lambda}$ is the number of corrugation periods fabricated onto the device and L is the device length. To minimise $\Delta\lambda_B$, we require large N (or L) and small Λ. Furthermore, the resulting Q-factor depends directly on N, such that $Q = \frac{\lambda_B}{\Delta\lambda_B} \propto N$. This trait is unique to DFB lasers and differs from Fabry–Perot VCSELs and WGM lasing cavities.

In most cases, the poor thermal conductivity of Perovskite active layers are partially circumvented through the use of heat sink capping layers such as sapphire [124, 125]. For Perovskite's visible lasing output, the first Bragg order ($m = 1$) imposes a strict requirement for $\Lambda < 200$ nm, which is extremely difficult to achieve through regular electron beam lithography or ion-milling methods. Therefore, the second or third Bragg orders ($m = 2, 3$) are much easier to achieve and demonstrate in reports [125]. A complete set of characterisations on Perovskite DFB lasing with varying Λ has been reported by Saliba et al. [126].

Jia et al. was the first to report a second Bragg ordered (TE polarised) lasing in $CH_3NH_3PbI_3$ based DFBs, as shown by the double-lobed far-field output in Fig. 3.30a [124]. In addition, the term "lasing death" (Fig. 3.30b) was coined after observing the dissipation of lasing intensities at 160 K after ~25 ns into quasi-CW excitation turn-on [124]. Interestingly, through studying the behaviour of high energy PL tail with varying pump pulse duration, they extracted an almost unchanging carrier temperature, which indicates the lasing death does not originate from insufficient heat dissipation [124]. Instead, they speculated this phenomenon to originate from

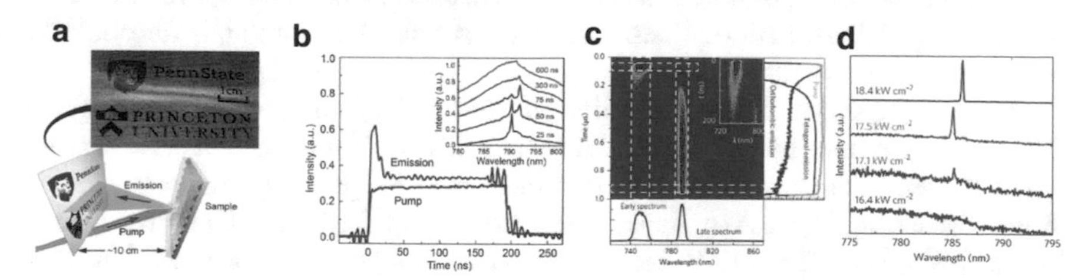

Fig. 3.30 Single mode lasing in $CH_3NH_3PbI_3$ based DFBs. **a** Demonstration of $m = 2$ DFB lasing, where a doubled-lobed far-field output was observed [124]. **b** PL kinetics showing lasing death of DFB under 195 ns pulsed excitation at 10.9 kWcm^{-2} and 2 kHz repetition rate at 160 K [124]. Subsequent demonstration of CW-pumped lasing in $CH_3NH_3PbI_3$ based DFBs by **c** employing mixed orthorhombic-tetragonal phases for engineering energy funnelling for optical buildup operating at 100 K. **d** CW pump intensity dependent spectra of DFB lasing at 100 K

photoinduced increase in dielectric constant in response to structural distortion. As a result, it steers the primary carriers from excitons to free carriers, quenching the initial exciton-mediated gain [124]. In the next report, Jia et al. proceeded to demonstrate CW-pumped lasing in $CH_3NH_3PbI_3$ based DFBs by manipulating the population inversion buildup through mixed Orthorhombic(host)-Tetragonal(localised) phases in the excitation volume (Fig. 3.30c, d) [125]. Here, the normally existing orthorhombic (larger gap) phase acts as the host while the tetragonal phase (lower gap) is structurally induced and kinetically trapped within the excitation volume as a result of local heating effects [125]. Essentially, the carrier sink effect from orthorhombic to tetragonal phase mimics a quasi-four level system that can efficiently produce low threshold lasing. Figure 3.30c shows the streak camera image from a $CH_3NH_3PbI_3$ PTF at 106 K, showing an ASE signature occurring at spectral positions expected of $CH_3NH_3PbI_3$ in tetragonal phase instead of orthorhombic phase. Figure 3.30d shows the CW-pump dependent single mode DFB lasing at 106 K, revealing a CW-pumping threshold of ~17kWcm^{-2} that is sustainable for up to ~1 h [125].

Recently, nano-imprinted lithographic (NIL) method has been proposed to upscale production of Perovskite based DFB lasers [127–129], although alternatives such as electron-beam [130] & self-healing [131] lithographic and inkjet printing [132] methods have also been suggested. By capitalising on the structural softness of Perovskites such as $CH_3NH_3PbI_3$, NIL has been shown viable for achieving CW-pumped DFB lasing [133] and 2D-photonic crystal DFB lasing [129]. From a manufacturing standpoint, NIL is a simple, high throughput and low cost process involving the use of a hard mould to transfer its patterns onto the Perovskite layer physically, as shown in Fig. 3.31a [133]. Figure 3.31b, c shows the room temperature CW pumped lasing spectra and its corresponding sustained duration. In this work, Li et al. reported an ultralow threshold of ~13Wcm^{-2} (Q ~ 1150) with an operation duration ~250 s for its single TE_0 mode [133]. Here, the surprisingly low room-temperature threshold is attributed to superior NIL fabrication technique coupled with improved spontaneous emission factor after NIL [133]. Figure 3.31d–f shows a similar NIL approach but applied in 2D-DFBs via photonic crystals patterning. Here, a chosen patterning corrugation and width of 450 and 230 nm gives rise to first-Bragg order, as seen by the inset's Kikuchi fringe lines. Using a ~ps pulsed excitation, a threshold of $\sim 3.8\ \mu$Jcm^{-2} is reported.

Figure 3.31f shows the manipulation of DFB lasing mode output by pumping different spots of the photonic crystal DFB. This effect arises due to spatially varying n_{eff} in the Perovskite layer thickness from the corrugated patterns. For completeness of the discussion, Fig. 3.31g shows the difference in DOP in NIL-DFBs against bare $CH_3NH_3PbI_3$ PTFs [127]. Here, Brenner et al. reports a clear transition from ASE's unpolarised output to strong linear polarised DFB output [127]. Interested readers may consult references [134] and [135] for other reports of 2D-photonic crystal DFB lasing in Perovskites. Another interesting work reported by Kim et al. involves the exploration of Perovskite DFB lasers fabricated using conventional LED-stacks, albeit with solution NIL grating patterning [136]. Under ps-pulsed excitation, lasing in both bottom- and top-emitting devices on glass and Si substrates showed

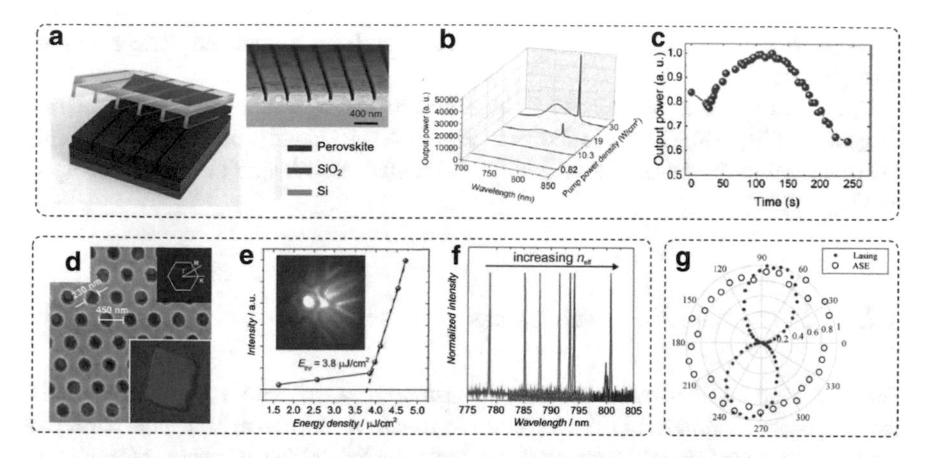

Fig. 3.31 Overview of Perovskite DFBs fabricated via nanoimprinted techniques for $CH_3NH_3PbI_3$. **a** Schematic patterning of DFB corrugation on PTF supported by SEM image [133]. **b** Room temperature CW-pump dependent lasing spectra with **c** sustained outputs up to ~250 s [133]. **d** SEM image of NIL 2D-photonic crystal patterning with inset showing the sample under oblique white light illumination [129]. **e** Fluence dependent lasing spectra with inset showing far-field patterns with Kikuchi lines [129]. **f** Fine-tunable DFB laser output upon variation of excitation site of photonic crystal DFB [129]. Comparison of DOP in Perovskite ASE and NIL-DFB lasing [127]

low threshold lasing in room temperature with peak EQE ~ 0.1% but showed large leakage currents when excited above $J = 2\,\text{Acm}^{-2}$ [136]. Kim et al. acknowledges that while such low current injection does not seem to adversely affect optically pumped lasing thresholds, efforts are required to eliminate these shunt channels by making smoother Perovskite films with smaller areas and impedance-matched device designs in order to reach the 2 Acm^{-2} limit in the ~ns pulsed excitation regime [136].

Outside of conventional $CH_3NH_3PbX_3$ based DFB lasers, Roh et al. reported an interesting class of DFB compatible lasers made of mixed cationic system $Cs_xMA_{1-x}PbI_{3-3x}Br_{3x}$ treated with additive 4-fluorobenzylammonium iodide bromide ($FPMAI_{1-x}Br_x$). Figure 3.32a shows the spectral tuning of DFB outputs

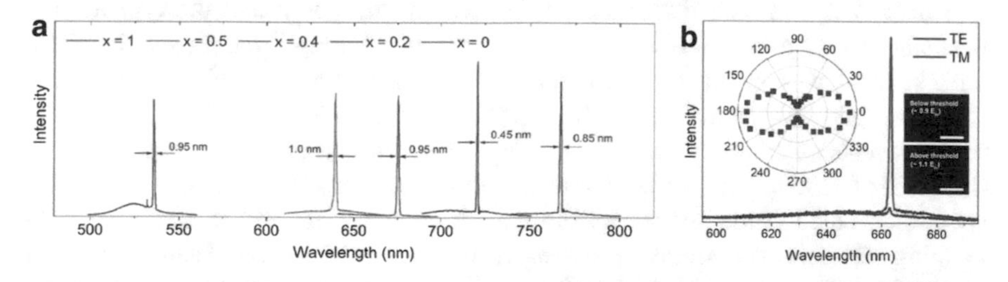

Fig. 3.32 DFB lasing characterisation in $FPMAI_{1-x}Br_x$ treated $Cs_xMA_{1-x}PbI_{3-3x}Br_{3x}$ DFBs over a range of suitable corrugation periods for each DFB [137]. **a** Spectral tuning via changing values of x in composition and **b** strongly TE polarised lasing output in $FPMAI_{0.8}Br_{0.2}$ – $Cs_{0.2}MA_{0.8}PbI_{2.4}Br_{0.6}$ DFBs. Inset shows the near-field image of lasing output below and above threshold excitation (Scale bar: 500 μm)

of varying compositional values of x with carefully calculated Bragg grating corrugations [137].

Specifically, for $x = 0.2$, The single-mode lasing output of $FPMAI_{0.8}Br_{0.2}$-$Cs_{0.2}MA_{0.8}PbI_{2.4}Br_{0.6}$ DFBs is shown in Fig. 3.32b. Evidently, the output is strongly TE polarised under ps-pulsed excitation, with sustained lasing operation for up to ~42 h [137].

3.2.4 Perovskite Random Lasers

The last kind of Perovskite laser is the random laser, where Perovskite media are purposely disordered [138] in order to create a morphological landscape that promotes random optical feedback through elastic Rayleigh scattering mechanisms. Practical applications of random lasing are yet to be seen but have been speculated to be applicable in speckle-free laser imaging and biomedical diagnosis. Importantly, the claim of random lasing should be supported by satisfying the Anderson's localisation condition (ALC):

$$k_p\langle l\rangle < 1 \tag{3.38}$$

which expresses that the mean-free-path of photons must be lesser than its optical cycle of $1/k_p$. Random lasing is stochastic in nature and does not have gain selectivity mechanisms, unlike the previous lasing mechanisms discussed. Thus, fluctuation in random lasing intensities and spectral position can be expected. Based on the random resonator statistical framework by Apalkov and Raikh, the lasing threshold I_{th} as a function of pump spot size is given by [139]:

$$I_{\text{th}} \propto e^{-\left(\frac{\left(\ln\frac{A}{A_0}\right)}{G}\right)^{1/\lambda}} \tag{3.39}$$

where A and A_0 are the illumination areas of the pump and the typical area occupied by a quasi-mode while G is the disorder strength parameter. In 2014, Dhanker et al. was the first to report random lasing phenomenon in highly disordered $CH_3NH_3PbI_3$ PMC networks in 2014 [139]. As shown in Fig. 3.33a–c, the SEM image reveals an extremely rough surface profile, with an estimated random lasing threshold of ~195 μJcm^{-2} [139]. Under a pumping spot of ~140 μm, bright scattering spots distributed randomly across the excitation area observed in spatio-spectral imaging suggests random lasing arising from efficient site-to-site scattering based waveguiding fortuitously linked in closed-loop paths. In addition, Dhanker et al. proposed that the dependence of random lasing threshold with illumination spot area via Eq. (3.39) is undeniable evidence of highly correlated network disorder with gain competition leading to chaotic lasing intensity variations [139]. By approximating

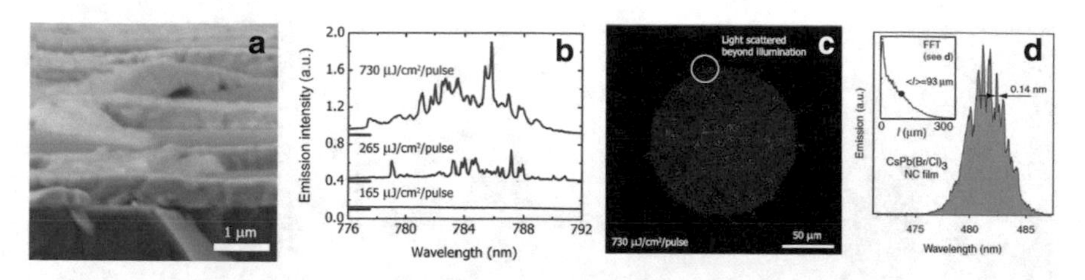

Fig. 3.33 Random lasing in highly disordered PTFs. **a** Cross-sectional SEM image of $CH_3NH_3PbI_3$ PTFs [139]. **b** Lasing spectra collected below, near and above threshold and **c** its microscopic image under a 140 μm spot excitation, showing random lasing from randomly illuminated scattering sites [139]. **d** First demonstration of Random lasing in $CsPb(Cl/Br)_3$ PNC films, with inset showing mean-free path l averaged over 25 pump laser shots [106]

Eq. (3.39) along the $\lambda \rightarrow 1$ limit, the intermediate value of $\frac{1}{G} \sim 0.7$ can be used to retrofit data relating random lasing threshold with illumination spot area to get the true disorder parameter of $\lambda \sim 1.9$ [139]. This indicated correlated disorder that supports small amounts of quasi-modes that has characteristic lengths of $\langle l \rangle \sim 50\,\mu$m [139]. Similar conclusions of random lasing also in $CH_3NH_3PbI_3$ PTFs were discussed by Shi et al., which also showed similar values of $\frac{1}{G} \sim 0.67$ [140]. Reduced random lasing thresholds through the use of patterned substrates have also been reported [141].

Figure 3.33d shows the random lasing spectrum for one of the pump laser shots, on a PNC film with several micron thickness [106]. While the lasing modes are stochastic, a statistical distribution comprising of 256 consecutive shots determined the average mean-free path length $l \sim 93\,\mu$m [106]. However, verification with the ALC reveals the product $k_p \langle l \rangle \sim 10^3 \gg 1$, and was argued to be a case of weakly scattered random lasing. On the basis that random lasing thrives in agglomerating or self-assembled PNC ensembles, single and two-photon pumped random lasing in ~100–200 nm sized $CH_3NH_3PbBr_3$ PNCs have also been reported [142]. Interestingly, three-photon pumped (3PP) random lasing in $CH_3NH_3PbBr_3$ PTFs was reported, although this assignment is debatable due to its spectral resemblance to a regular ASE profile [143].

Another report of random lasing was shown in structural two-dimensional FA-(N-MPDA)$PbBr_4$ PMWs by Roy et al., as shown in Fig. 3.34 [144]. Figure 3.34a shows the optical image of an as-synthesized FA-(N-MPDA)$PbBr_4$ PMW with rod length ~3 mm. The inset shows the corresponding top-view SEM image. Figure 3.34b shows the angular dependent lasing spectra of the PMW. Here, the purpose was to verify the stochastic nature of random lasing, where relative variation in pump angle can showcase different quasi-mode outputs. By tilting the PMW within a wide range of $\pm 20°$, lasing outputs, albeit with varying spectral modes were still observed, indicating the multi-directionality and randomness of the output [144]. Figure 3.34c, d shows the fluence dependent random lasing spectra under a 55 ps pulsed excitation. Here, the threshold was determined at ~500 nJcm^{-2}, with an estimated Q-factor of ~5350. Finally, Fig. 3.34e shows that under the ps-pulsed excitation conditions, random lasing in the PMW can be sustained up to ~2 h.

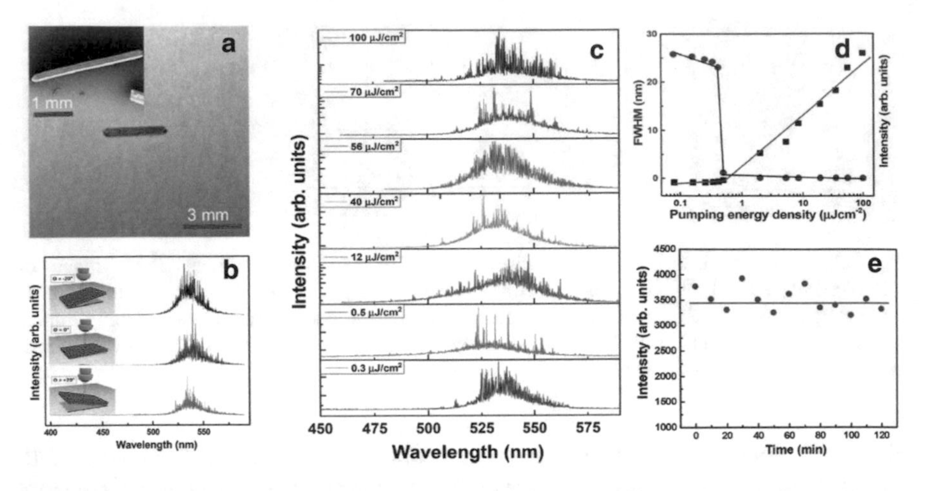

Fig. 3.34 Random lasing characterisation in FA-(N-MPDA)$PbBr_4$ PMWs [144]. **a** Optical image of the as-synthesized PMW with ~3 mm length. Inset shows the top-view SEM image of PMW. **b** Angular dependence of random lasing within ±20° under fixed pumping fluence of 12 μJcm^{-2}. **c** Fluence dependent lasing spectra at 0°, showing **d** a threshold feature along 500nJcm^{-2} with a simultaneous FWHM narrowing. **e** Random lasing stability of PMW pumped at 12 μJcm^{-2}

3.3 Summary and Conclusions

In this chapter, we examined the various optical gain mechanisms of common Perovskite media by focusing on its carrier dynamics. Importantly, detailed knowledge of its carrier optical buildup and avalanche stimulated emission energy scheme provides insights for engineering lower pump threshold Perovskite lasers. In Sects. 3.1.1, 3.1.2, 3.1.3, 3.1.4 and 3.1.5, we discussed the various photonic gain arising from carrier buildups in free-carriers, biexcitons, trions and single excitons. Primarily, the assignments were done using steady-state PL observations and supported by time-resolved PL and absorption kinetics. In Sect. 3.1.6, we also discussed another type of amplification mechanism, called the polariton gain which is commonly found in highly anisotropic Perovskite media with large oscillator strengths. In this case, the optical gain arises from the polaritonic condensate macroscopic emission instead of carrier population inversion buildups followed by avalanche stimulated emission. For this reason, polariton lasing is extremely appealing due to its much lower gain threshold than regular photonic lasing means. Next, in Sect. 3.2, we focused on addressing the characterisation and fabrication techniques for various Perovskite cavity configurations. In the case of Perovskite VCSELs, meticulous efforts to ensure parallel alignment of the top DBR with active layer and the bottom DBR together with minimisation of undesirable air columns. In the case of DFB lasers, the crucial factor to consider is to align the grating's Bragg-wavelength $\lambda_B \sim \lambda_{lase}$. Furthermore, Single-mode operation can be promoted through using wider surfaces L so that more corrugations N can be printed. As a result of Perovskites' visible range emission, reasonable fabrication for DFB lasing is only planned for second Bragg-ordered outputs in order to maintain low costs of manufacturing these devices. Next, we also discussed the various interesting research highlights in Perovskite WGM lasing microcavities. Here, frequency upconversion up to three-photon regime has been demonstrated via the passivation of active PMCs with hBN. In addition, the so-called vernier effect serves as an excellent way to extract single-mode WGM lasing in coupled differently sized Perovskite microstructure for high precision optical sensing applications. Lastly, we discussed that random lasing is mostly observed in highly disordered PTFs, characterised by their large disorder parameter G. From Eq. 3.39, a value of $\frac{1}{G} \sim 0.67$ affirms random lasing occuring from highly correlated disorder that supports small number of quasi-modes under the premise of quasi-three dimensional random model.

References

1. S.A. Veldhuis et al., Benzyl alcohol-treated CH3NH3PbBr 3 nanocrystals exhibiting high luminescence, stability, and ultralow amplified spontaneous emission thresholds. Nano Lett. **17**(12), 7424–7432 (2017)
2. Y. Wang, X. Li, J. Song, L. Xiao, H. Zeng, H. Sun, All-inorganic colloidal perovskite quantum dots: a new class of lasing materials with favorable characteristics. Adv. Mater. **27**(44), 7101–7108 (2015)
3. H. Chosrowjan, Fluorescence up-conversion Methods and Applications, in *Encyclopedia of Spectroscopy and Spectrometry (Third Edition)*. ed. by J.C. Lindon, G.E. Tranter, D.W. Koppenaal (Academic Press, Oxford, 2017), pp. 654–660
4. V. Klimov et al., Optical gain and stimulated emission in nanocrystal quantum dots. Science 290(5490), 314–317 (2000)
5. V.I. Klimov et al., Single-exciton optical gain in semiconductor nanocrystals. Nature **447**(7143), 441–446 (2007)
6. Y. Wang, M. Zhi, Y.-Q. Chang, J.-P. Zhang, Y. Chan, Stable, ultralow threshold amplified spontaneous emission from CsPbBr 3 nanoparticles exhibiting trion gain. Nano Lett. **18**(8), 4976–4984 (2018)
7. M. Duguay, J.-W. Hansen, An ultrafast light gate. Appl. Phys. Lett. **15**(6), 192–194 (1969)
8. K. Chen, A.J. Barker, F.L. Morgan, J.E. Halpert, J.M. Hodgkiss, Effect of carrier thermalization dynamics on light emission and amplification in organometal halide perovskites. J. Phys. Chem. Lett. **6**(1), 153–158 (2015)
9. W. Zhao et al., Optical gain from biexcitons in CsPbBr 3 nanocrystals revealed by two-dimensional electronic spectroscopy. J. Phys. Chem. Lett. **10**(6), 1251–1258 (2019)
10. S.D. Stranks, V.M. Burlakov, T. Leijtens, J.M. Ball, A. Goriely, H.J. Snaith, Recombination kinetics in organic-inorganic perovskites: excitons, free charge, and subgap states. Phys. Rev. Appl. **2**(3), 034007 (2014)
11. F. Deschler et al., High photoluminescence efficiency and optically pumped lasing in solution-processed mixed halide perovskite semiconductors. J. Phys. Chem. Lett. **5**(8), 1421–1426 (2014)
12. J.S. Manser, P.V. Kamat, Band filling with free charge carriers in organometal halide perovskites. Nat. Photonics **8**(9), 737–743 (2014)
13. S.W. Eaton et al., Lasing in robust cesium lead halide perovskite nanowires. Proc. Natl. Acad. Sci. **113**(8), 1993–1998 (2016)
14. X. Wang et al., Cesium lead halide perovskite triangular nanorods as high-gain medium and effective cavities for multiphoton-pumped lasing. Nano Res. **10**(10), 3385–3395 (2017)

15. P. Geiregat et al., Using bulk-like nanocrystals to probe intrinsic optical gain characteristics of inorganic lead halide perovskites. ACS Nano **12**(10), 10178–10188 (2018)
16. B. Tang et al., Single-mode lasers based on cesium lead halide perovskite submicron spheres. ACS Nano **11**(11), 10681–10688 (2017)
17. M. Ghanassi, M. Schanne-Klein, F. Hache, A. Ekimov, D. Ricard, C. Flytzanis, Time-resolved measurements of carrier recombination in experimental semiconductor-doped glasses: confirmation of the role of Auger recombination. Appl. Phys. Lett. **62**(1), 78–80 (1993)
18. K. Asano, T. Yoshioka, Exciton–Mott physics in two-dimensional electron–hole systems: phase diagram and single-particle spectra. J. Phys. Soc. Jpn. **83**(8), 084702 (2014)
19. G. Xing et al., Low-temperature solution-processed wavelength-tunable perovskites for lasing. Nat. Mater. **13**(5), 476–480 (2014)
20. K. Wu, Y.-S. Park, J. Lim, V.I. Klimov, Towards zero-threshold optical gain using charged semiconductor quantum dots. Nat. Nanotechnol. **12**(12), 1140–1147 (2017)
21. K. Wei, X. Zheng, X. Cheng, C. Shen, T. Jiang, Observation of ultrafast exciton–exciton annihilation in CsPbBr 3 quantum dots. Adv. Opt. Mater. **4**(12), 1993–1997 (2016)
22. G. Delport et al., Exciton–exciton annihilation in two-dimensional halide perovskites at room temperature. J. Phys. Chem. Lett. **10**(17), 5153–5159 (2019)
23. G. Denton, N. Tessler, N. Harrison, R. Friend, Factors influencing stimulated emission from poly (p-phenylenevinylene). Phys. Rev. Lett. **78**(4), 733 (1997)
24. Y. Yu, Y. Yu, C. Xu, A. Barrette, K. Gundogdu, L. Cao, Fundamental limits of exciton-exciton annihilation for light emission in transition metal dichalcogenide monolayers. Phys. Rev. B **93**(20), 201111 (2016)
25. M.A. Lampert, Mobile and immobile effective-mass-particle complexes in nonmetallic solids. Phys. Rev. Lett. **1**(12), 450 (1958)
26. K. Kheng, R. Cox, M.Y. d'Aubigné, F. Bassani, K. Saminadayar, S. Tatarenko, Observation of negatively charged excitons X−in semiconductor quantum wells. Phys. Rev. Lett. **71**(11), 1752 (1993)
27. A. Singh et al., Trion formation dynamics in monolayer transition metal dichalcogenides. Phys. Rev. B **93**(4), 041401 (2016)
28. F. Jakubka, S.B. Grimm, Y. Zakharko, F. Gannott, J. Zaumseil, Trion electroluminescence from semiconducting carbon nanotubes. ACS Nano **8**(8), 8477–8486 (2014)
29. S.M. Santos et al., All-optical trion generation in single-walled carbon nanotubes. Phys. Rev. Lett. **107**(18), 187401 (2011)
30. E.V. Shornikova et al., Negatively charged excitons in CdSe nanoplatelets. Nano Lett. **20**(2), 1370–1377 (2020)
31. N. Yarita et al., Observation of positive and negative trions in organic-inorganic hybrid perovskite nanocrystals. Phys. Rev. Mater. **2**(11), 116003 (2018)
32. L. Peng et al., Bright trion emission from semiconductor nanoplatelets. Phys. Rev. Mater. **4**(5), 056006 (2020)
33. S. Rosen, O. Schwartz, D. Oron, Transient fluorescence of the off state in blinking CdSe/CdS/ZnS semiconductor nanocrystals is not governed by Auger recombination. Phys. Rev. Lett. **104**(15), 157404 (2010)
34. D.E. Gómez, J. Van Embden, P. Mulvaney, M.J. Fernée, H. Rubinsztein-Dunlop, Exciton−trion transitions in single CdSe–CdS core–shell nanocrystals. ACS Nano **3**(8), 2281–2287 (2009)
35. J. Zhao, G. Nair, B.R. Fisher, M. G. Bawendi, Challenge to the charging model of semiconductor-nanocrystal fluorescence intermittency from off-state quantum yields and multiexciton blinking. Phys. Rev. Lett. **104**(15), 157403 (2010)
36. I.Y. Eremchev, I. Osad'ko, A. Naumov, Auger ionization and tunneling neutralization of single CdSe/ZnS nanocrystals revealed by excitation intensity variation. J. Phys. Chem. C **120**(38), 22004–22011 (2016)
37. J.A. Castañeda et al., Efficient biexciton interaction in perovskite quantum dots under weak and strong confinement. ACS Nano **10**(9), 8603–8609 (2016)

38. S. Nakahara et al., Suppression of trion formation in CsPbBr 3 perovskite nanocrystals by postsynthetic surface modification. J. Phys. Chem. C **122**(38), 22188–22193 (2018)
39. Y. Kanemitsu, Trion dynamics in lead halide perovskite nanocrystals. J. Chem. Phys. **151**(17), 170902 (2019)
40. N.S. Makarov, S. Guo, O. Isaienko, W. Liu, I. Robel, V.I. Klimov, Spectral and dynamical properties of single excitons, biexcitons, and trions in cesium–lead-halide perovskite quantum dots. Nano Lett. **16**(4), 2349–2362 (2016)
41. F. Hu et al., Slow Auger recombination of charged excitons in nonblinking perovskite nanocrystals without spectral diffusion. Nano Lett. **16**(10), 6425–6430 (2016)
42. N. Yarita et al., Impact of postsynthetic surface modification on photoluminescence intermittency in formamidinium lead bromide perovskite nanocrystals. J. Phys. Chem. Lett. **8**(24), 6041–6047 (2017)
43. S. Das, S. Kallatt, N. Abraham, K. Majumdar, Gate-tunable trion switch for excitonic device applications. Phys. Rev. B **101**(8), 081413 (2020)
44. R.-B. Liu, W. Yao, L. Sham, Quantum computing by optical control of electron spins. Adv. Phys. **59**(5), 703–802 (2010)
45. H.C. Wang et al., High-performance CsPb1−xSnxBr 3 perovskite quantum dots for light-emitting diodes. Angew. Chem. **129**(44), 13838–13842 (2017)
46. J. Puls, G. Mikhailov, F. Henneberger, D.R. Yakovlev, A. Waag, Optical gain and lasing of trions in delta-doped ZnSe quantum wells, in *10th International Symposium on Nanostructures: Physics and Technology*, vol. 5023 (International Society for Optics and Photonics, 2003), pp. 376–378
47. V.I. Klimov, A.A. Mikhailovsky, D. McBranch, C.A. Leatherdale, M.G. Bawendi, Quantization of multiparticle Auger rates in semiconductor quantum dots. Science **287**(5455), 1011–1013 (2000)
48. H. Kwon et al., Probing trions at chemically tailored trapping defects. ACS Cent. Sci. **5**(11), 1786–1794 (2019)
49. C. Dang, J. Lee, C. Breen, J.S. Steckel, S. Coe-Sullivan, A. Nurmikko, Red, green and blue lasing enabled by single-exciton gain in colloidal quantum dot films. Nat. Nanotechnol. **7**(5), 335–339 (2012)
50. C. Grivas et al., Single-mode tunable laser emission in the single-exciton regime from colloidal nanocrystals. Nat. Commun. **4**(1), 1–9 (2013)
51. G.H. Ahmed, J. Yin, O.M. Bakr, O.F. Mohammed, Successes and challenges of core/shell lead halide perovskite nanocrystals. ACS Energy Lett. **6**(4), 1340–1357 (2021)
52. L. Zhao et al., Low-threshold stimulated emission in perovskite quantum dots: single-exciton optical gain induced by surface plasmon polaritons at room temperature. J. Mater. Chem. C **8**(17), 5847–5855 (2020)
53. S. Chen, A. Nurmikko, Excitonic gain and laser emission from mixed-cation halide perovskite thin films. Optica **5**(9), 1141–1149 (2018)
54. J. Navarro-Arenas, I. Suárez, V.S. Chirvony, A.F. Gualdrón-Reyes, I. Mora-Seró, J. Martínez-Pastor, Single-exciton amplified spontaneous emission in thin films of CsPbX3 (X= Br, I) perovskite nanocrystals. J. Phys. Chem. Lett. **10**(20), 6389–6398 (2019)
55. J. Shi et al., Low-threshold stimulated emission of hybrid perovskites at room temperature through defect-mediated bound excitons. arXiv preprint arXiv:1902.07371 (2019)
56. T. Byrnes, N.Y. Kim, Y. Yamamoto, Exciton–polariton condensates. Nat. Phys. **10**(11), 803–813 (2014)
57. A.P. Schlaus, M.S. Spencer, X. Zhu, Light–matter interaction and lasing in lead halide perovskites. Acc. Chem. Res. **52**(10), 2950–2959 (2019)
58. M. Baranowski et al., Exciton binding energy and effective mass of CsPbCl 3: a magneto-optical study. Photonics Res. **8**(10), A50–A55 (2020)
59. J.D. Plumhof, T. Stöferle, L. Mai, U. Scherf, R.F. Mahrt, Room-temperature Bose-Einstein condensation of cavity exciton–polaritons in a polymer. Nat. Mater. **13**(3), 247–252 (2014)
60. H. Deng, G. Weihs, D. Snoke, J. Bloch, Y. Yamamoto, Polariton lasing vs. photon lasing in a semiconductor microcavity. Proc. Natl. Acad. Sci. **100**(26), 15318–15323 (2003)

61. R. Su et al., Room-temperature polariton lasing in all-inorganic perovskite nanoplatelets. Nano Lett. **17**(6), 3982–3988 (2017)
62. A. Imamog, R. Ram, S. Pau, Y. Yamamoto, Nonequilibrium condensates and lasers without inversion: exciton-polariton lasers. Phys. Rev. A **53**(6), 4250 (1996)
63. S. Luo, Y. Wang, L. Liao, Z. Zhang, X. Shen, Z. Chen, Exciton-polariton dynamics modulated by exciton-photon detuning in a ZnO microwire. J. Appl. Phys. **127**(2), 025702 (2020)
64. T.-C. Lu et al., Room temperature polariton lasing vs. photon lasing in a ZnO-based hybrid microcavity. Opt. Express **20**(5), 5530–5537 (2012)
65. G. Lanty et al., Hybrid cavity polaritons in a ZnO-perovskite microcavity. Phys. Rev. B **84**(19), 195449 (2011)
66. E. Kammann, H. Ohadi, M. Maragkou, A.V. Kavokin, P.G. Lagoudakis, Crossover from photon to exciton-polariton lasing. New J. Phys. **14**(10), 105003 (2012)
67. J. Wu et al., Polariton lasing in InGaN quantum wells at room temperature. Opto-Electron. Adv. **2**(12), 190014 (2019)
68. W. Deng, X. Jin, Y. Lv, X. Zhang, X. Zhang, J. Jie, 2D Ruddlesden-popper perovskite nanoplate based deep-blue light-emitting diodes for light communication. Adv. Func. Mater. **29**(40), 1903861 (2019)
69. H. Deng, G. Weihs, C. Santori, J. Bloch, Y. Yamamoto, Condensation of semiconductor microcavity exciton polaritons. Science **298**(5591), 199 (2002). https://doi.org/10.1126/science.1074464
70. S. Christopoulos et al., Room-temperature polariton lasing in semiconductor microcavities. Phys. Rev. Lett. **98**(12), 126405 (2007). https://doi.org/10.1103/PhysRevLett.98.126405
71. T. Guillet et al., Polariton lasing in a hybrid bulk ZnO microcavity. Appl. Phys. Lett. **99**(16), 161104 (2011). https://doi.org/10.1063/1.3650268
72. J.D. Plumhof, T. Stoferle, L. Mai, U. Scherf, R.F. Mahrt, Room-temperature Bose-Einstein condensation of cavity exciton-polaritons in a polymer. Nat. Mater. **13**(3), 247–252 (2014). https://doi.org/10.1038/nmat3825
73. S. Kéna-Cohen, S.R. Forrest, Room-temperature polariton lasing in an organic single-crystal microcavity. Nat. Photon. **4**(6), 371–375 (201). https://doi.org/10.1038/nphoton.2010.86
74. K.S. Daskalakis, S.A. Maier, R. Murray, S. Kena-Cohen, Nonlinear interactions in an organic polariton condensate. Nat Mater. **13**(3), 271–278 (2014). https://doi.org/10.1038/nmat3874
75. A.P. Schlaus et al., How lasing happens in CsPbBr 3 perovskite nanowires. Nat. Commun. **10**(1), 1–8 (2019)
76. H. Dong, C. Zhang, X. Liu, J. Yao, Y.S. Zhao, Materials chemistry and engineering in metal halide perovskite lasers. Chem. Soc. Rev. **49**(3), 951–982 (2020)
77. T.J. Evans et al., Continuous-wave lasing in cesium lead bromide perovskite nanowires. Adv. Opt. Mater. **6**(2), 1700982 (2018)
78. J. Kasprzak et al., Bose–Einstein condensation of exciton polaritons. Nature **443**(7110), 409–414 (2006)
79. F.P. Laussy, G. Malpuech, A. Kavokin, Spontaneous coherence buildup in a polariton laser. Phys. Status Solidi (C) **1**(6), 1339–1350 (2004)
80. D. Porras, C. Tejedor, Linewidth of a polariton laser: theoretical analysis of self-interaction effects. Phys. Rev. B **67**(16), 161310 (2003)
81. J. Kasprzak et al., Bose-Einstein condensation of exciton polaritons. Nature **443**(7110), 409–414 (2006). https://doi.org/10.1038/nature05131
82. X. Wang et al., High-quality in-plane aligned CsPbX3 perovskite nanowire lasers with composition-dependent strong exciton–photon coupling. ACS Nano **12**(6), 6170–6178 (2018)
83. Q. Shang et al., Role of the exciton–polariton in a continuous-wave optically pumped CsPbBr 3 perovskite laser. Nano Lett. **20**(9), 6636–6643 (2020)
84. Q. Han et al., Transition between exciton-polariton and coherent photonic lasing in all-inorganic perovskite microcuboid. ACS Photonics **7**(2), 454–462 (2020)
85. S. Zhang et al., Strong exciton–photon coupling in hybrid inorganic–organic perovskite micro/nanowires. Adv. Opt. Mater. **6**(2), 1701032 (2018)

86. L. Polimeno et al., Observation of two thresholds leading to polariton condensation in 2D hybrid perovskites. Adv. Opt. Mater. **8**(16), 2000176 (2020)
87. K. Gauthron et al., Optical spectroscopy of two-dimensional layered (C6H5C2H4-NH3) 2-PbI4 perovskite. Opt. Express **18**(6), 5912–5919 (2010)
88. T. E. Sale, *Vertical Cavity Surface Emitting Lasers* (Research Studies Press, 1995)
89. T. Yoshikawa, H. Kosaka, K. Kurihara, M. Kajita, Y. Sugimoto, K. Kasahara, Complete polarization control of 8 $\times$ 8 vertical-cavity surface-emitting laser matrix arrays. Appl. Phys. Lett. **66**(8), 908–910 (1995)
90. Y. Wang, X. Li, V. Nalla, H. Zeng, H. Sun, Solution-processed low threshold vertical cavity surface emitting lasers from all-inorganic perovskite nanocrystals. Adv. Func. Mater. **27**(13), 1605088 (2017)
91. U. W. Pohl, *Epitaxy of Semiconductors: Physics and Fabrication of Heterostructures* (Springer Nature, 2020)
92. Y.-Y. Liu, T.-C. Wu, P.S. Yeh, Etch-stop process for precisely controlling the vertical cavity length of GaN-based devices. Mater. Sci. Semicond. Process. **120**, 105265 (2020)
93. T. Akagi et al., High-quality AlInN/GaN distributed Bragg reflectors grown by metalorganic vapor phase epitaxy. Appl. Phys. Express **13**(12), 125504 (2020)
94. Y. Xiao et al., Single-nanowire single-mode laser. Nano Lett. **11**(3), 1122–1126 (2011)
95. C. Zhao et al., High performance single-mode vertical cavity surface emitting lasers based on CsPbBr3 nanocrystals with simplified processing. Chem. Eng. J. 127660 (2020)
96. P.D. Cunningham, J.o.B. Souza Jr, I. Fedin, C. She, B. Lee, D.V. Talapin, Assessment of anisotropic semiconductor nanorod and nanoplatelet heterostructures with polarized emission for liquid crystal display technology. ACS Nano **10**(6), 5769–5781 (2016)
97. Y. Mei et al., Quantum dot vertical-cavity surface-emitting lasers covering the 'green gap'. Light Sci. Appl. **6**(1), e16199 (2017)
98. S. Chen, C. Zhang, J. Lee, J. Han, A. Nurmikko, High-Q, low-threshold monolithic perovskite thin-film vertical-cavity lasers. Adv. Mater. **29**(16), 1604781 (2017)
99. M.S. Alias, Z. Liu, A. Al-Atawi, T.K. Ng, T. Wu, B.S. Ooi, Continuous-wave optically pumped green perovskite vertical-cavity surface-emitter. Opt. Lett. **42**(18), 3618–3621 (2017)
100. C. Tian, S. Zhao, W. Zhai, C. Ge, G. Ran, Low-threshold room-temperature continuous-wave optical lasing of single-crystalline perovskite in a distributed reflector microcavity. RSC Adv. **9**(62), 35984–35989 (2019)
101. Z. Gu et al., Two-photon pumped CH3NH3PbBr 3 perovskite microwire lasers. Adv. Opt. Mater. **4**(3), 472–479 (2016)
102. F. Sasaki, H. Mochizuki, Y. Zhou, Y. Sonoda, R. Azumi, Optical pumped lasing in solution processed perovskite semiconducting materials: Self-assembled microdisk lasing. Jpn. J. Appl. Phys. **55**(4S), 04ES02 (2016)
103. K. Wang, W. Sun, J. Li, Z. Gu, S. Xiao, Q. Song, Unidirectional lasing emissions from CH3NH3PbBr 3 perovskite microdisks. ACS Photonics **3**(6), 1125–1130 (2016)
104. W. Zhang et al., Controlling the cavity structures of two-photon-pumped perovskite microlasers. Adv. Mater. **28**(21), 4040–4046 (2016)
105. N. Kurahashi, V.-C. Nguyen, F. Sasaki, H. Yanagi, Whispering gallery mode lasing in lead halide perovskite crystals grown in microcapillary. Appl. Phys. Lett. **113**(1), 011107 (2018)
106. S. Yakunin et al., Low-threshold amplified spontaneous emission and lasing from colloidal nanocrystals of caesium lead halide perovskites. Nat. Commun. **6**(1), 1–9 (2015)
107. Q. Zhang, S.T. Ha, X. Liu, T.C. Sum, Q. Xiong, Room-temperature near-infrared high-Q perovskite whispering-gallery planar nanolasers. Nano Lett. **14**(10), 5995–6001 (2014)
108. Q. Zhang, R. Su, X. Liu, J. Xing, T.C. Sum, Q. Xiong, High-quality whispering-gallery-mode lasing from cesium lead halide perovskite nanoplatelets. Adv. Func. Mater. **26**(34), 6238–6245 (2016)
109. Q. Liao, K. Hu, H. Zhang, X. Wang, J. Yao, H. Fu, Perovskite microdisk microlasers self-assembled from solution. Adv. Mater. **27**(22), 3405–3410 (2015)
110. G. Li et al., Record-low-threshold lasers based on atomically smooth triangular nanoplatelet perovskite. Adv. Func. Mater. **29**(2), 1805553 (2019)

111. B.R. Sutherland et al., Perovskite thin films via atomic layer deposition. Adv. Mater. **27**(1), 53–58 (2015)
112. M.B. Price et al., Whispering-gallery mode lasing in perovskite nanocrystals chemically bound to silicon dioxide microspheres. J. Phys. Chem. Lett. **11**(17), 7009–7014 (2020)
113. M. Ghulinyan, D. Navarro-Urrios, A. Pitanti, A. Lui, G. Pucker, L. Pavesi, Whispering-gallery modes and light emission from a Si-nanocrystal-based single microdisk resonator. Opt. Express **16**(17), 13218–13224 (2008)
114. I. Teraoka, S. Arnold, Enhancing the sensitivity of a whispering-gallery mode microsphere sensor by a high-refractive-index surface layer. JOSA B **23**(7), 1434–1441 (2006)
115. C. Lam, P.T. Leung, K. Young, Explicit asymptotic formulas for the positions, widths, and strengths of resonances in Mie scattering. JOSA B **9**(9), 1585–1592 (1992)
116. M. Gomilšek, M. Ravnik, Whispering gallery modes. Univ. Ljubl. Ljubl. Semin **11** (2011)
117. X. Lopez-Yglesias, J.M. Gamba, R.C. Flagan, The physics of extreme sensitivity in whispering gallery mode optical biosensors. J. Appl. Phys. **111**(8), 084701 (2012)
118. Q. Li et al., Lasing from reduced dimensional perovskite microplatelets: Fabry-Pérot or whispering-gallery-mode? J. Chem. Phys. **151**(21), 211101 (2019)
119. K. Wang, Z. Gu, S. Liu, J. Li, S. Xiao, Q. Song, Formation of single-mode laser in transverse plane of perovskite microwire via micromanipulation. Opt. Lett. **41**(3), 555–558 (2016)
120. Z. Shi et al., Robust frequency-upconversion lasing operated at 400 K from inorganic perovskites microcavity. Nano Res. 1–10 (2021)
121. H. Yu et al., Waterproof perovskite-hexagonal boron nitride hybrid nanolasers with low lasing thresholds and high operating temperature. ACS Photonics **5**(11), 4520–4528 (2018)
122. B. Zhou et al., Linearly polarized lasing based on coupled perovskite microspheres. Nanoscale **12**(10), 5805–5811 (2020)
123. N. Pourdavoud, Stimulated emission and lasing in metal halide perovskites by direct thermal nanoimprint. Universität Wuppertal, Fakultät für Elektrotechnik, Informationstechnik und ..., (2020)
124. Y. Jia, R.A. Kerner, A.J. Grede, A.N. Brigeman, B.P. Rand, N.C. Giebink, Diode-pumped organo-lead halide perovskite lasing in a metal-clad distributed feedback resonator. Nano Lett. **16**(7), 4624–4629 (2016)
125. Y. Jia, R.A. Kerner, A.J. Grede, B.P. Rand, N.C. Giebink, Continuous-wave lasing in an organic–inorganic lead halide perovskite semiconductor. Nat. Photonics **11**(12), 784–788 (2017)
126. M. Saliba et al., Structured organic–inorganic perovskite toward a distributed feedback laser. Adv. Mater. **28**(5), 923–929 (2016)
127. P. Brenner et al., Highly stable solution processed metal-halide perovskite lasers on nanoimprinted distributed feedback structures. Appl. Phys. Lett. **109**(14), 141106 (2016)
128. G.L. Whitworth et al., Nanoimprinted distributed feedback lasers of solution processed hybrid perovskites. Opt. Express **24**(21), 23677–23684 (2016)
129. N. Pourdavoud et al., Photonic nanostructures patterned by thermal nanoimprint directly into organo-metal halide perovskites. Adv. Mater. **29**(12), 1605003 (2017)
130. K. Wang et al., High-density and uniform lead halide perovskite nanolaser array on silicon. J. Phys. Chem. Lett. **7**(13), 2549–2555 (2016)
131. D. Xing et al., Self-healing lithographic patterning of perovskite nanocrystals for large-area single-mode laser array. Adv. Func. Mater. **31**(1), 2006283 (2021)
132. F. Mathies, P. Brenner, G. Hernandez-Sosa, I.A. Howard, U.W. Paetzold, U. Lemmer, Inkjet-printed perovskite distributed feedback lasers. Opt. Express **26**(2), A144–A152 (2018)
133. Z. Li et al., Room-temperature continuous-wave operation of organometal halide perovskite lasers. ACS Nano **12**(11), 10968–10976 (2018)
134. S. Chen et al., A photonic crystal laser from solution based organo-lead iodide perovskite thin films. ACS Nano **10**(4), 3959–3967 (2016)
135. J. Moon et al., Environmentally stable room temperature continuous wave lasing in defect-passivated perovskite. arXiv preprint arXiv:1909.10097 (2019)

136. H. Kim et al., Optically pumped lasing from hybrid perovskite light-emitting diodes. Adv. Opt. Mater. **8**(1), 1901297 (2020)
137. K. Roh et al., Widely tunable, room temperature, single-mode lasing operation from mixed-halide perovskite thin films. ACS Photonics **6**(12), 3331–3337 (2019)
138. A. Safdar, Y. Wang, T.F. Krauss, Random lasing in uniform perovskite thin films. Opt. Express **26**(2), A75–A84 (2018)
139. R. Dhanker, A. Brigeman, A. Larsen, R. Stewart, J.B. Asbury, N.C. Giebink, Random lasing in organo-lead halide perovskite microcrystal networks. Appl. Phys. Lett. **105**(15), 151112 (2014)
140. Z.-F. Shi et al., Near-infrared random lasing realized in a perovskite CH 3 NH 3 PbI 3 thin film. J. Mater. Chem. C **4**(36), 8373–8379 (2016)
141. G. Weng et al., Giant reduction of the random lasing threshold in CH 3 NH 3 PbBr 3 perovskite thin films by using a patterned sapphire substrate. Nanoscale **11**(22), 10636–10645 (2019)
142. S. Liu et al., Random lasing actions in self-assembled perovskite nanoparticles. Opt. Eng. **55**(5), 057102 (2016)
143. G. Weng et al., Picosecond random lasing based on three-photon absorption in organometallic halide CH3NH3PbBr 3 perovskite thin films. ACS Photonics **5**(7), 2951–2959 (2018)
144. P.K. Roy et al., Unprecedented random lasing in 2D organolead halide single-crystalline perovskite microrods. Nanoscale **12**(35), 18269–18277 (2020)

Chapter 4
Outlook and Conclusions

4.1 General Conclusions

In this book, we provided extensive discussions on key highlights in the Perovskite laser research field. In Chap. 1, we begin by briefly introducing the concept of stimulated emission, as it is key to the coherent properties of lasers. We have shown that minimally population inversion of carriers in three-level gain materials is a must for optical gain. Upon the inclusion of an external cavity, stringent boundary conditions impose further selection of allowed photon modes that can participate in laser mode oscillations. Importantly, while single mode laser outputs are preferred due to spectral purity, they require narrower cavity lengths (larger FSR) and may arbitrarily increase the gain threshold, and thus, pumping threshold. Lastly, we briefly discussed the two degrees of optical coherence of lasers, known as the temporal (monochromaticity) and spatial (directionality) coherence. We have also distinguished lasing from ASE based on the differences in their light amplification mechanisms. In Chap. 2, we briefly introduce the generic crystal structures of perovskites in varying structural dimensionalities. In principle, the structural stability and phase depends on the mutual compatibility of A, B and X ions as given by the octahedral and Goldschmidt tolerance factors. Next, we gave an overview of the achievement milestones over the last 7 years and presented the on-going research interests and development for Perovskite lasers. As proof of versatility and potential of the Perovskites gain media, we have also provided an overview of the multitude of ASE and lasing reports for a variety of Perovskite morphologies, ranging from bulk single-crystals (PSCs) down to nanocrystals (PNCs). Interestingly, for Perovskite microstructures, lasing commonly occurs as a consequence of morphological waveguiding effects of the optical modes, where optical feedback occurs between the end facets without requiring any external reflective surfaces. In Chap. 3, we focused on the light amplification mechanisms of Perovskites. Generally, the knowledge of basic dynamics and interactions of a material's primary photogenerated carriers (e.g., excitons or free carriers) is essential when

Y. K. E. Tay et al., *Halide Perovskite Lasers*, Nanoscience and Nanotechnology,
https://doi.org/10.1007/978-981-16-7973-5_4

attempting to study the underlying carrier dynamics leading to optical gain (mechanism). Lead iodide Perovskites generally possess the lowest exciton binding energies in room temperature and so, its optical gain mechanisms are straightforwardly assigned to free electron and hole carriers (free-carriers). On the other hand, biexcitonic gain and trionic gain mechanisms have been proposed for primarily excitonic $CH_3NH_3PbBr_3$ and $CsPbBr_3$ systems, where single exciton gain mechanisms cannot occur naturally due to its two-level system analogue. However, we also included a discussion on reports claiming to observe single exciton gain in mixed-cationic Perovskite thin-films. For completeness, we also considered another form of lasing, called the polariton lasing. Polariton carriers are generally found in morphologically anisotropic materials with large oscillator strengths and unlike photon lasing, polariton lasing does not require population inversion. Next, we moved on to discuss several strategies adopted by researchers in reducing optical losses while boosting stimulated emission in various laser cavity configurations. Importantly, nano-imprint and inkjet printed Perovskite laser arrays hold the potential of synthetic upscaling suitable for industrial manufacturing and commercialisation. Finally, we provided an overview on the current state-of-the-art Perovskite lasers, where we highlighted various strategies adopted in reports, to gradually increase the thermal stability of Perovskite lasers under longer durational pulsed excitations that gravitate towards CW pumping. This chapter completes this book by listing the remaining challenges that stands between electrically pumped lead halide Perovskite lasers and its projected environmental impact.

4.2 Remaining Challenges of Perovskite Lasers

Thus far, optical spectroscopic studies of Perovskite lasers have revealed many evidence of its potential in producing sustainable light energy, based on merits such as low costs of fabrication and production, sub-μJ cm^{-2} pumping thresholds and relatively high gain coefficients along $\sim 10^3$cm^{-1}. However, several key issues remain largely unexplored and requires attention before Perovskite lasers could make it to the commercialisation stage. In this section, we shall consider strategies for handling the toxicity of lead-based Perovskite media and its thermal stability issues that hinders its progress towards electrical pumping regime [1].

4.2.1 Toxicity of Lead Halide Perovskite Based Lasers

Lead precursors and solvents (DMF, DMSO and toluene) used in synthesizing Perovskites are ecologically toxic [1–3]. Furthermore, the pursuit of device performance optimisation and large-scale production strategies often results in further manipulation of solvent chemistry, which adds to the aggregated ecological impact to the environment [1]. Such toxic chemicals are classified as “substances of very

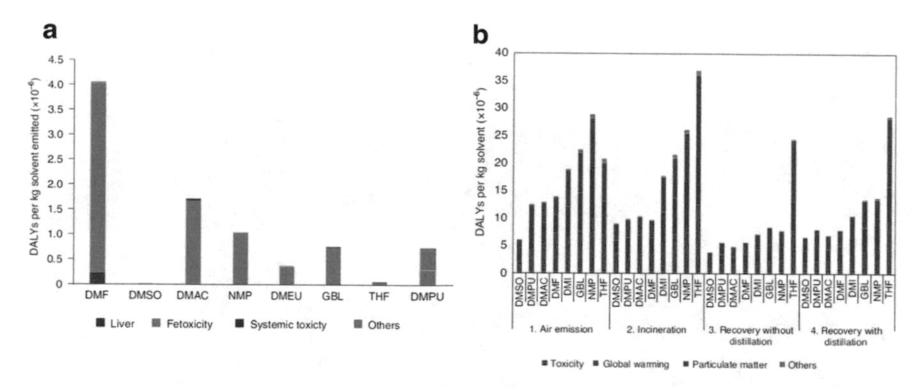

Fig. 4.1 The Disability-Adjusted Life Year (DALY) analysis [1, 4] for **a** various health impacts caused by various common aprotic solvents and **b** the various ecological impacts

high concern" (SVHC) by the European Chemical Agency and the gradual replacement with ecologically safer substances is highly encouraged. Thus, the lead halide Perovskite lasers are expected to face obstacles to commercialisation, where stringent laws or protocols must be adhered for safe-handling and disposal [4]. Towards this goal, comprehensive Life-Cycle Analysis (LCA) studies on the long-term effects of solvents, electrodes, charge selective transport materials and Perovskite precursors on both human health and environmental impacts must be carried out [1, 4]. Figure 4.1a shows the Disability-Adjusted Life Year (DALY) analysis revealed that the commonly used DMF solvent is most lethal and that other solvents such as NMP and GBL generally showed trends of causing fetotoxicity, thereby threatening human reproduction. Thus, this suggests that a set of safety guidelines is necessary for people working in the solvent production, Perovskite syntheses labs and operating the Perovskite based devices. Interestingly, Fig. 4.1b showed that while DMF is extremely toxic to human health, its environmental footprint in terms of toxicity, global warming and particulate pollution are much less harmful than GBL, NMP and THF.

Other sources of environmental pollution in the production of Perovskite lasers are lead ions, its metallic end-electrodes and charge-selective transport layers, all of which form the end product device [4]. Contrary to popular belief (Fig. 4.2), a study of B-cations Perovskite solar cells conducted by Serrano et al. found that lead ions (Pb^{2+}) generally pose little health and environmental impacts and are in fact lower toxicity than tin ions (Sn^{2+}). Instead, the study point to the solar cell's back electrodes (e.g. Au) as the major culprit behind human toxicity (cancerous and non-cancerous), freshwater eutrophication & ecotoxicities, and non-renewable depletion [4]. Since the device stack of solar cells is analogous to lasers, similar conclusions should apply in this case. Thus, scientists and manufacturers intending to commercialise lead-based Perovskite lasers should be aware of the inaccuracy of the preconceived notion of lead as the major source of toxicity and pollution [4].

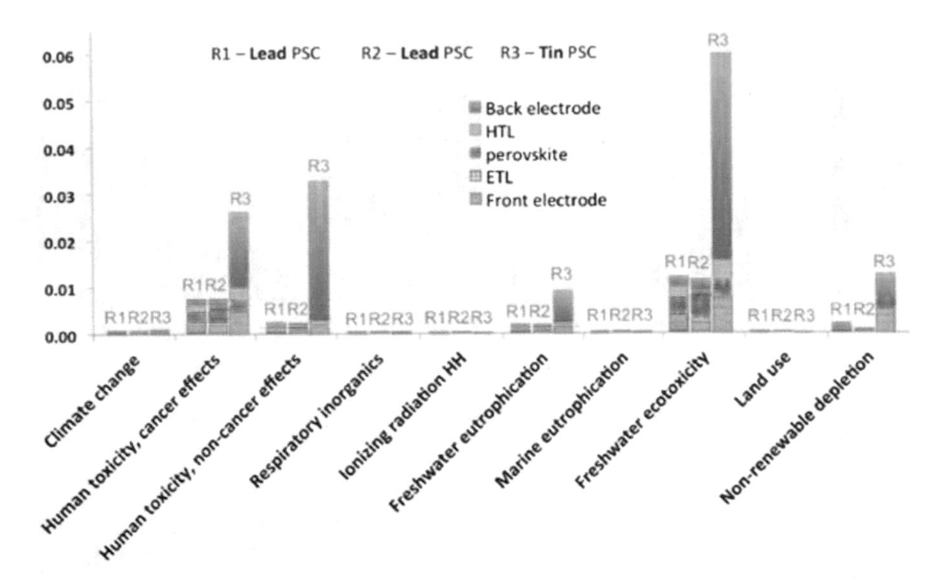

Fig. 4.2 The comparison of Normalised Environmental impact values between lead-based and tin-based Perovskite solar cells [4]. Evidently, the major source of health and environmental impact is surprisingly the metallic back-end electrodes rather than lead ions

With sufficient knowledge of the health and environmental impact of lead-based Perovskite laser devices, the last part of the LCA is to discuss its "end-of-life" strategies, as shown in Fig. 4.3. In Fig. 4.3a, incineration of Perovskite solar cells (and lasers) is encouraged while landfill disposal is highly discouraged, mainly due to health threats such as cancerous, non-cancerous toxicity and marine eutrophication [4]. In fact, landfill disposal is extremely unsuitable, as bioaccumulation of lead and other toxic chemicals can leak into the groundwater and infiltrate agriculture

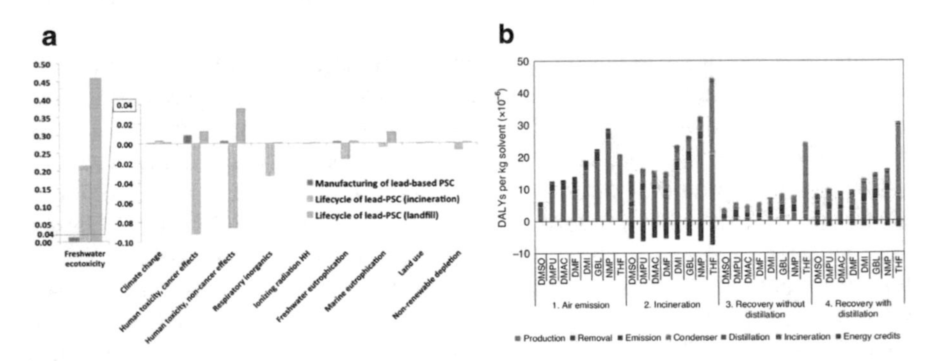

Fig. 4.3 **a** The normalised environmental and health impact values of manufacturing and disposing lead-based Perovskite solar-cells (PSCs) [4]. **b** The DALY analysis for various end-of-life solvent disposal/retrieval strategies, with both incineration and solvent recovery with distillation yielding energy credits [1]

[2, 3]. Other health consequences include mental retardation and developmental delay in young children, as well as congenital paralysis, chronic nervous system and other intestinal organ damages in adults [5, 6]. Similar conclusion encouraging incineration of Perovskite devices at its end-of-life are shown in Fig. 4.3b, where incineration and distillation recovery of solvents can lead to accumulation of energy credits [1]. Note that additional costs may be incurred for recovery strategies, such as post-treatment waste handling and generation of side products (e.g. polyethylene terephthalate, which can cause water intoxication [4]) and emissions incurred for instance, in waste incineration and distillation [1]. Figure 4.4 illustrates the typical life-cycle of Perovskite devices.

With reference to Figs. 4.1, 4.2, and 4.3, it is evident that both the production and disposal ends of lead-based Perovskite devices entailed serious health and environmental impacts, it is clear that safety and environmental protection protocols must be drafted, covering (i) manufacturing raw substances and Perovskite device fabrication and (ii) end-of-life disposal, re-cycling and solvent recovery are needed [7]. As such, it can be projected that the future commercial viability of lead Perovskite laser devices will depend greatly on strict laws or until safer alternatives (solvents and electrode choices) can be found, while maintaining lasing device efficiencies.

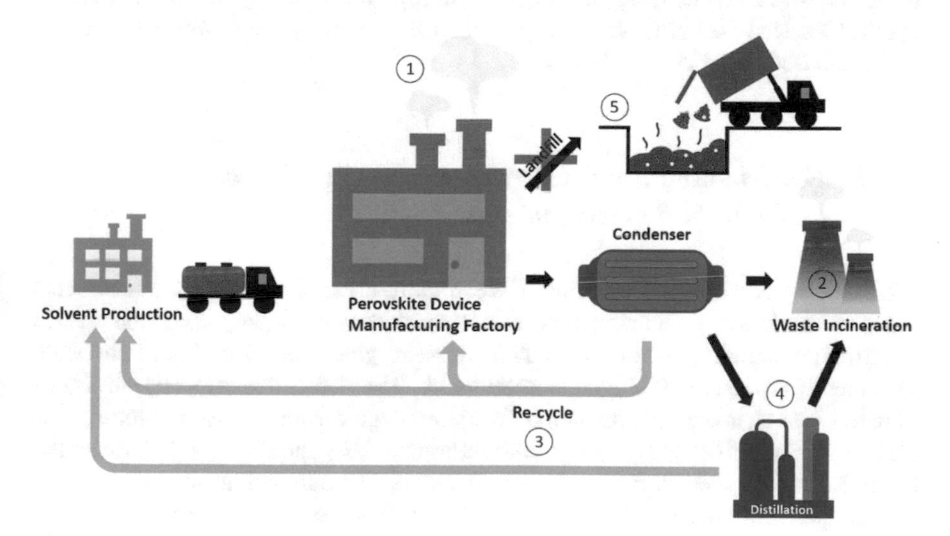

Fig. 4.4 A life-cycle schematic illustrating the production and disposing of lead-based Perovskite devices. Solvents are produced and transported to factories fabricating Perovskite solar cells and laser devices. Subsequently, the end-of-life strategies proposed are either (1) direct solvent emission to the environment or by first solvent condensation and (2) waste incineration, (3) direct re-cycling of condensed solvents or (4) distillation for solvent and precursor recovery, of which any other remaining undistilled matter are incinerated. An undesirable disposal is (5) landfill, as the bioaccumulation of lead and other toxic solvents will pollute agriculture and freshwater that is critical to human sustenance and survival

Currently, less toxic routes for fabricating lead halide Perovskites have been proposed [8, 9], while others focus on substituting lead with other group 14 (IV-A) metals, although they generally suffer from decreased stability in maintaining the 2+ oxidation states in when moving up in the periodic table [10–12]. For instance, the active oxidation interplay between the Sn^{2+} and Sn^{4+} states causes destabilisation of Perovskite structures and had to be stabilised by SnF_2 treatment for stable lasing applications [13]. Alternatively, rare-earth substitutes may also be considered, as the Europium-based $CsEuCl_3$ PNCs have been synthesized and reported to possess very similar optoelectronic properties to its lead-counterparts [9]. This work is exciting for reasons appealing beyond toxicity but also as an alternative promising candidate for developing stable deep-blue Perovskite lasers, which is still currently a huge research gap in the field. Conventional $CsPbCl_3$ and $CH_3NH_3PbCl_3$ had faced problems in producing room-temperature [14] deep-blue ASE and lasing due to strong competition of non-radiative channels arising from defect states [15]. For the convenience of readers who are interested in understanding the development of other lead-free (bismuth, germanium and etc.-based) Perovskites, we encourage them to refer to a review by Hang et al. [16] and Yi et al. [17].

However, the stability of the 2+ oxidation state decreases when moving up within the group 14 (IV-A) elements (tin and germanium). The efficiency and stability values reached for lead-free or tin-based approaches are currently much worse than those for lead-based PSC [15, 18, 19, 22].

4.2.2 Electrically Pumped Perovskite Lasing: Device and Thermal Managements

While the successful demonstration of electrically-driven Perovskite lasers remains a major challenge, optimising both the Perovskite optical properties and device construction should gradually lead us to this end goal. For an optimal Perovskite gain media, we require (I) high quantum yields (low defect density), (II) slow non-radiative loss channels (in population inversion, large carrier densities regime), (III) high carrier mobilities and (IV) high thermal conductivity and stability [18]. Recently, Perovskites have seen various success with stable CW pumped lasing, as attributed to high quantum yields with low defect density, suppression of multiparticle auger recombination and high carrier mobility. However, **maximisation of carrier densities** as well as device engineering tailored at material **thermal stability** are major hurdles that remains to be tackled. In other words, scientists should focus efforts on device structural engineering aimed at improving electrical injection efficiencies and reducing device heat generation under prolonged device operation by introducing compact and feasible heatsink mechanisms for complementing Perovskite active layers [19–21].

Carrier densities maximisation in response to fixed electrical injections is important as it indicates a lower injection threshold is required for achieving population

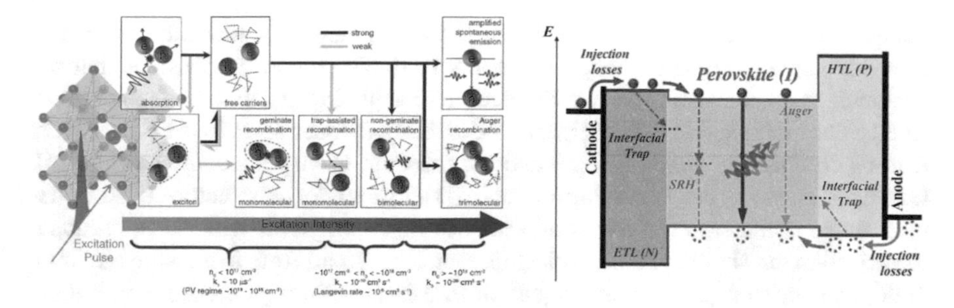

Fig. 4.5 Illustration of all photophysical processes occurring in a free-carrier based $CH_3NH_3PbI_3$ thin film system under the influence of increasing carrier densities as a result of increasing excitation densities

inversion. In a device stack surrounding the Perovskite active layer, its performance efficiency greatly depends on the interplay between several dynamical processes shown in Fig. 4.5a. Here, for instance, bimolecular ASE buildup in a free carrier systems such as $CH_3NH_3PbI_3$ often competes with non-radiative (trimolecular or higher) Auger recombination and is known to raise optical pumping threshold from 10^{17} to $10^{18}cm^{-3}$ in $CH_3NH_3PbI_3$ PTFs [22]. For devices, electrical injection losses due to interfacial traps found between charge transporting to active layers (Fig. 4.5b) may also contribute to raising the device's operating injection threshold. By developing strategies to suppress auger recombination while increasing stimulated emission recombination rates and reducing injection losses, one can arbitrarily lower the injection thresholds and increase the net modal gain coefficients. Essentially, the maximisation of injected carrier density can be done by optimising a Perovskite active layer's quantum yield.

Given that k_1, 2, 3 are the mono-, bi- and trimolecular recombination rates and n to be the corresponding carrier density, the quantum yields Φ can be defined as:

$$\Phi_{FC}(n) = \frac{nk_2}{nk_2 + \left(k_1 + n^2k_3\right)} \tag{4.1}$$

$$\Phi_X(n) = \frac{k_1}{k_1 + nk_2} \tag{4.2}$$

In a free-carrier system described by Eq. (4.1), the carrier density refers to the electron and hole carrier densities such that $n = n_e = n_h$. Evidently, increasing the active layer's Φ_{FC} reduces the trap densities ($\downarrow k_1$) and auger recombination ($\downarrow k_3$). Similarly, for excitonic systems, the trionic and biexcitonic carrier maximisation firstly depends on the maximisation of their primary exciton densities. As described by Eq. (4.2), increasing the active layer's Φ_X reduces the auger (X–X annihilation) recombination ($\downarrow k_2$) channel, thereby increasing the efficiency of trionic or biexcitonic carrier formation and build-ups. Recently, as proof-of-concept, Yao et al. reported a significant increase in $CsPbBr_3$ PNC based LEDs' external quantum efficiencies (EQE) from 5.5 to 9.1% coupled with low efficiency roll-offs, via surface ligand exchange strategies [23]. The report emphasizes that the exchange from conventional long-chained oleic acid/oleylamine ligands to cerium-tributylphosphine oxide hybrid ligands effectively reduces surface traps on PNCs, leading to quantum yield increase and suppression of auger recombination [23]. Given the similarity between a Perovskite LED and laser device structure, such strategies aimed at increasing electroluminescence quantum yields will aid in reducing the injection threshold required for lasing operation. Apart from increasing quantum yield, another strategy to reduce injection threshold is through a change in optical gain mechanism. For excitonic systems, trionic gain is favored over biexciton gain, as the former relatively mitigates auger X–X annihilation along the sub-excitonic carrier density regime [24, 25]. Near-zero optically pumped thresholds for doubly charged inorganic Chalcogenide QDs have been observed [24, 25] and are expected to translate to equivalently low electrical injection thresholds. However, these efforts are yet to be seen in PNCs, possibly because of its complex surface chemistry involved and therefore remains as a research gap.

A common mechanism for severe heating in Perovskite devices is due to auger-heating caused by auger recombination. As such, effective thermal management strategies should be integrated into the resulting device in order to complement the Perovskite active layer's poor thermal conductivity. For comparison, lead halide Perovskites $\left(k_3 \sim 3 \times 10^{-28} \mathrm{cm}^6\, \mathrm{s}^{-1}\right)$ are found to possess one order of magnitude higher auger recombination rates than GaAs $\left(k_3 \sim 2 \times 10^{-27} \mathrm{cm}^6\, \mathrm{s}^{-1}\right)$. At the same time, studies also revealed that slower auger rates in tin-based $FASnI_3$ PTFs $\left(k_3 \sim 9.3 \times 10^{-27} \mathrm{cm}^6\, \mathrm{s}^{-1}\right)$ [26], despite its slightly higher toxicity. While this perspective seem to showcase a better suitability of tin-based Perovskites as better gain media than their lead-based counterparts, the former is also plagued by severe stability issues that relies on SnF_2 treatment [13]. In a review by Sargent et al. [18], a simulation (Fig. 4.6a) assuming near unity quantum yield Perovskite

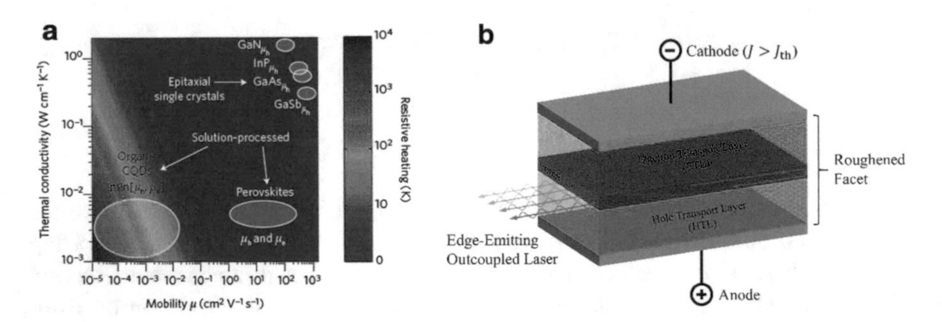

Fig. 4.6 **a** A simulation of the Perovskite thermal conductivity in comparison to organics and epitaxial single crystal semiconductors [18]. The simulation was run under the assumption of complete radiative recombination in a 200 nm thick Perovskite film under the injection density of 1 kA cm^{-2}, with a free-carrier density of 10^{18} cm^{-3} generated. **b** Schematic of a proposed Perovskite laser diode consisting of end electrodes, followed by charge selective transport layers sandwiching a central Perovskite active layer

material with 200 nm film thickness possesses thermal conductance of $<0.3 \text{ W m}^{-1} \text{K}^{-1}$ under an injection density of 1 kA cm^{-2} and free-carrier density of 10^{18} cm^{-3}. In real Perovskite systems with non-negligible Auger recombination, its true thermal conductivity is even lower than this simulated value. In other words, thermal management should be achieved by introducing additional cooling elements in the device stack, such as the use of Al_2O_3 layer for heat-sink effects to overcome lasing deaths in a DFB device structure [27].

Finally, the achievement of electrically driven Perovskite laser devices ends with a proper combination of layers complementing the Perovskite active layer. Good knowledge of layer planning and strategies used to tackle efficiency roll-offs could serve to benefit in this aspect, as Perovskite lasers are similar to its LED (PeLED) counterparts, except that the former operates in much higher current injection regimes $\left(J > 100 \text{ A cm}^{-2}\right)$ and the latter at much lower current injections $\left(J < 1 \text{ A cm}^{-2}\right)$ [28].

With the achievement of EQE > 20% in PeLEDs, the likelihood of developing Perovskite laser diodes is fairly reasonable, as intrinsically long carrier diffusion lengths and mobilities [29], coupled with negligible Auger losses were reported for PeLEDs [28]. To project potential issues faced by a laser diode designed similar to a PeLED, scientists began exploring PeLEDs pumped under intense current densities [28, 30]. Interestingly, while meticulously designed "small-area" PeLEDs have shown tolerance and sustain operations at $J > 600 \text{ A cm}^{-2}$ [28] and $J > 900 \text{ A cm}^{-2}$ [30], both designs suffered from EQE roll-off due to Joule-heating and charge injection imbalances (instead of Auger losses) [28, 30].

Generally, the efficiency roll-off in Perovskite LEDs (and presumably in Perovskite laser diodes too) are caused by **(I) unbalanced charge injection**, **(II) Auger recombination (non-radiative)** and **(III) Joule-heating** [31]. In 2015, Yuan et al. reported that unbalanced charge injection would severely affect excitonic based PeLED systems as the net electric field created within the Perovskite layer internally, would result in exciton ionization and cause ASE quenching [32]. Electric field strengths up to 10^5 V cm^{-1} can greatly lower EQE and that careful selection is needed when coupling ETLs and HTLs to the end-electrodes. Very recently, it was reported that these three roll-off factors can be collectively mitigated in a single PeLED stack, by designing (i) energetic ladder for balanced charge transports, (ii) material optimisation by replacing insulating long chains of PEABr with ionic KBr that helps to reduce Auger recombination rates and (iii) replacement of glass with thermally dissipative substrates such as sapphire in a quasi-two dimensional Perovskite system [31]. Importantly, this work highlights that the resulting device can operate under a sustained current injection densities close to ~1 kA cm^{-2}, which is close to the Perovskite lasing injection regime [31]. In addition to replacement of glass with thermally dissipative sapphire substrate, heat sinks, small area injection via nanopatterning was shown to effectively combat Joule-heating by allowing the heat to dissipate to the surroundings. A combination of both passive and active cooling approaches may be necessary. Interestingly, this work acknowledges that despite the triple optimisation of the Perovskite LED, based on the figure of merit $J \times$ EQE calculations, it was determined that even their best LED sample was falling short of the lasing threshold required by at least one order of magnitude. Despite various valuable lessons that can be picked up from the Perovskite LEDs, it is important to also consider that Perovskite lasers require an external cavity for optical mode confinement. Since most PeLEDs stacks do not offer optical confinement, ultralow optically pumped ASE thresholds along sub-μJ cm^{-2} could only be acquired in most of these papers discussed above.

An early proposal of the Perovskite laser diode is the conventional edge-emitting laser diode structure shown in Fig. 4.6b [18]. Here, the diode consists of a thin Perovskite active layer (orange) sandwiched between much thicker charge transport layers of lower refractive indices. In this way, the stimulated emission is effectively confined within the Perovskite active layer. The roughened end-facet re-directs approaching stimulated emission light back while the smooth end-facet acts as the output coupler for lasing emission. Although the design looks simplistic, selecting suitable charge transport (ETL and HTL) layers can be complex because it can affect the Perovskite layer's interfacial stability and carrier dynamics. In addition, the proposed structure has a lack of optical confinement mechanisms (mirrors and etc.) and mostly rely on end-facet reflection and refractive index contrast induced waveguiding effects. Furthermore, the lack of heat-dissipative mechanisms in the stack shown in Fig. 4.5a imply that Joule-heating would continue to be an existing problem unless a thermally more stable Perovskite system can be found. Very recently, Qin et al. reported stable room temperature CW-pumped lasing in quasi-two dimensional Perovskite films integrated into a DFB grating layer without any thermal management strategies [33]. Here, the success was attributed to the suppression of triplet excitons by injecting a triplet-quencher such as oxygen or organic molecules with lower triplet energy than that of the Perovskite frameworks [33]. Interestingly, they reasoned that the main cause of concern for CW-pumped lasing death phenomenon is likely due to the active singlet–triplet exciton state annihilation (STA) that impedes optical gain during optical excitations. This highlights a much more deep-seated issue with the optical gain mechanism of Perovskite materials rather than several device-related concerns that make them premature for electrical injection studies. Although not discussed within the work, this result may however indicate that the quasi-2D gain media may potentially have relatively higher thermal conductivities than its archetypal three-dimensional counterparts, making the former more viable as candidates for deriving electrically-driven Perovskite lasers in the future.

4.3 Concluding Remarks

While excellent optically pumped lasing performance projects great potential for electrically pumped Perovskite lasers, the task is still ardous because of unoptimized material quantum yields (needed to raise radiative rates and suppress auger rates) and poor intrinsic thermal conductivity of Perovskite active layers. Furthermore, due to similarities in PeLEDs and Perovskite lasers, one needs to be mindful of (I) efficiency roll-offs due to Joule-heating and unbalanced charge injection and (II) suitability of electrodes and charge transport layers to the Perovskite layer such that no interfacial complication arises. In addition, unexplored factors such as electrical injection damage threshold and stability may pose as future problems upon the realisation of the first prototype Perovskite laser diode. Despite the multitude of problems that makes electrically driven Perovskite lasers seem daunting, we believe that the intensive international research efforts hold promise in solving these issues and this ultimate goal may be achieved within this or the next decade.

References

1. W. Deng, X. Jin, Y. Lv, X. Zhang, X. Zhang, J. Jie, 2D Ruddlesden-Popper perovskite nanoplate based deep-blue light-emitting diodes for light communication. Adv. Func. Mater. **29**(40), 1903861 (2019)
2. A. Babayigit, H.-G. Boyen, B. Conings, Environment versus sustainable energy: the case of lead halide perovskite-based solar cells. MRS Energy Sustain. **5** (2018)
3. E.M. Hutter, R. Sangster, C. Testerink, B. Ehrler, C.M. Gommers, Metal halide perovskite toxicity effects on plants are caused by iodide ions. arXiv preprint arXiv:2012.06219 (2020)
4. L. Serrano-Lujan, N. Espinosa, T.T. Larsen-Olsen, J. Abad, A. Urbina, F.C. Krebs, Tin-and lead-based perovskite solar cells under scrutiny: an environmental perspective. Adv. Energy Mater. **5**(20), 1501119 (2015)
5. E. Meyer, D. Mutukwa, N. Zingwe, R. Taziwa, Lead-free halide double perovskites: a review of the structural, optical, and stability properties as well as their viability to replace lead halide perovskites. Metals **8**(9), 667 (2018)
6. A.H. Slavney, R.W. Smaha, I.C. Smith, A. Jaffe, D. Umeyama, H.I. Karunadasa, Chemical approaches to addressing the instability and toxicity of lead–halide perovskite absorbers. Inorg. Chem. **56**(1), 46–55 (2017)
7. N. Espinosa, L. Serrano-Luján, A. Urbina, F.C. Krebs, Solution and vapour deposited lead perovskite solar cells: ecotoxicity from a life cycle assessment perspective. Sol. Energy Mater. Sol. Cells **137**, 303–310 (2015)
8. M. Imran et al., Benzoyl halides as alternative precursors for the colloidal synthesis of lead-based halide perovskite nanocrystals. J. Am. Chem. Soc. **140**(7), 2656–2664 (2018)
9. J. Huang et al., Lead-free cesium europium halide perovskite nanocrystals. Nano Lett. **20**(5), 3734–3739 (2020)
10. F. Hao, C.C. Stoumpos, R.P. Chang, M.G. Kanatzidis, Anomalous band gap behavior in mixed Sn and Pb perovskites enables broadening of absorption spectrum in solar cells. J. Am. Chem. Soc. **136**(22), 8094–8099 (2014)
11. M.H. Kumar et al., Lead-free halide perovskite solar cells with high photocurrents realized through vacancy modulation. Adv. Mater. **26**(41), 7122–7127 (2014)
12. N.K. Noel et al., Lead-free organic–inorganic tin halide perovskites for photovoltaic applications. Energy Environ. Sci. **7**(9), 3061–3068 (2014)
13. G. Xing et al., Solution-processed tin-based perovskite for near-infrared lasing. Adv. Mater. **28**(37), 8191–8196 (2016)
14. A.A. Lohar, A. Shinde, R. Gahlaut, A. Sagdeo, S. Mahamuni, Enhanced photoluminescence and stimulated emission in $CsPbCl_3$ nanocrystals at low temperature. J. Phys. Chem. C **122**(43), 25014–25020 (2018)
15. R. Ahumada-Lazo et al., Emission properties and ultrafast carrier dynamics of $CsPbCl_3$ perovskite nanocrystals. J. Phys. Chem. C **123**(4), 2651–2657 (2019)
16. H. Hu, B. Dong, W. Zhang, Low-toxic metal halide perovskites: opportunities and future challenges. J. Mater. Chem. A **5**(23), 11436–11449 (2017)

17. Z. Yi, N.H. Ladi, X. Shai, H. Li, Y. Shen, M. Wang, Will organic–inorganic hybrid halide lead perovskites be eliminated from optoelectronic applications? Nanoscale Adv. **1**(4), 1276–1289 (2019)
18. B.R. Sutherland, E.H. Sargent, Perovskite photonic sources. Nat. Photon. **10**(5), 295 (2016)
19. H.Y. Dong, C.H. Zhang, X.L. Liu, J.N. Yao, Y.S. Zhao, Materials chemistry and engineering in metal halide perovskite lasers (in English). Chem. Soc. Rev. **49**(3), 951–982 (2020). https://doi.org/10.1039/c9cs00598f
20. W.B. Gunnarsson, B.P. Rand, Electrically driven lasing in metal halide perovskites: challenges and outlook (in English). APL Mater. **8**(3), (2020). Artn 030902. https://doi.org/10.1063/1.5143265
21. J. Qin, X.-K. Liu, C. Yin, F. Gao,Carrier dynamics and evaluation of lasing actions in halide perovskites. Trends Chem. **3**(1), 34–46 (2021). https://doi.org/10.1016/j.trechm.2020.10.010
22. G. Xing et al., Low-temperature solution-processed wavelength-tunable perovskites for lasing. Nat. Mater. **13**(5), 476–480 (2014)
23. J.-S. Yao et al., Suppressing Auger recombination in cesium lead bromide perovskite nanocrystal film for bright light-emitting diodes. J. Phys. Chem. Lett. **11**(21), 9371–9378 (2020)
24. K.F. Wu, Y.S. Park, J. Lim, V.I. Klimov, Towards zero-threshold optical gain using charged semiconductor quantum dots (in English). Nat. Nanotechnol. **12**(12), 1140+ (2017). https://doi.org/10.1038/Nnano.2017.189
25. O.V. Kozlov, Y.S. Park, J. Roh, I. Fedin, T. Nakotte, V.I. Klimov, Sub-single-exciton lasing using charged quantum dots coupled to a distributed feedback cavity (in English). Science **365**(6454), 672+ (2019). https://doi.org/10.1126/science.aax3489
26. R.L. Milot, G.E. Eperon, T. Green, H.J. Snaith, M.B. Johnston, L.M. Herz, Radiative monomolecular recombination boosts amplified spontaneous emission in $HC(NH_2)_2SnI_3$ perovskite films. J. Phys. Chem. Lett. **7**(20), 4178–4184 (2016)
27. Y. Jia, R.A. Kerner, A.J. Grede, B.P. Rand, N.C. Giebink, Continuous-wave lasing in an organic–inorganic lead halide perovskite semiconductor. Nat. Photon. **11**(12), 784–788 (2017)
28. H. Kim et al., Hybrid perovskite light emitting diodes under intense electrical excitation. Nat. Commun. **9**(1), 1–9 (2018)
29. G. Xing et al., Long-range balanced electron-and hole-transport lengths in organic-inorganic CH3NH3PbI3. Science **342**(6156), 344–347 (2013)
30. F. Yuan et al., All-inorganic hetero-structured cesium tin halide perovskite light-emitting diodes with current density over 900 A cm^{-2} and its amplified spontaneous emission behaviors. Physica Status Solidi (RRL) Rapid Res. Lett. **12**(5), 1800090 (2018)
31. C. Zou, Y. Liu, D.S. Ginger, L.Y. Lin, Suppressing efficiency roll-off at high current densities for ultra-bright green perovskite light-emitting diodes. ACS Nano **14**(5), 6076–6086 (2020)
32. F. Yuan et al., Electric field-modulated amplified spontaneous emission in organo-lead halide perovskite CH3NH3PbI3. Appl. Phys. Lett. **107**(26), 261106 (2015)
33. C. Qin et al., Stable room-temperature continuous-wave lasing in quasi-2D perovskite films. Nature **585**(7823), 53–57 (2020)

Printed in the United States
by Baker & Taylor Publisher Services